John Keast Lord, Francis Walker

A List of Hymenoptera

In Egypt, in the neighbourhood of the Red Sea, and in Arabia; with

description of the new species

John Keast Lord, Francis Walker

A List of Hymenoptera
In Egypt, in the neighbourhood of the Red Sea, and in Arabia; with description of the new species

ISBN/EAN: 9783744780414

Printed in Europe, USA, Canada, Australia, Japan

Cover: Foto ©berggeist007 / pixelio.de

More available books at **www.hansebooks.com**

A

LIST OF HYMENOPTERA

COLLECTED BY

J. K. LORD, ESQ.

IN EGYPT,

IN THE NEIGHBOURHOOD OF THE RED SEA,

AND IN ARABIA.

WITH DESCRIPTIONS OF THE NEW SPECIES.

BY

FRANCIS WALKER, F.L.S.

LONDON:

E. W. JANSON, 28, MUSEUM STREET, W.C.

1871.

LONDON:

E. NEWMAN, PRINTER, DEVONSHIRE STREET, BISHOPSGATE.

PREFACE.

—o—

THE aspect of these regions indicates that they are especially adapted to Aculeate Hymenoptera, and the following list shows that Mr. Lord has made an extensive collection of them, and that he has discovered numerous new species, though many of this tribe were published by Klug in the work which is here often quoted. Mr. Frederick Smith, of the British Museum, has kindly arranged the specimens and has named many of the species already described.

Very few of the species here enumerated are natives of North Europe; several of them inhabit the Mediterranean region; others have a wider range, and extend to South Africa and to West Africa, or to Hindostan. The other tribes of Hymenoptera are very slightly traced, excepting the Chrysidæ and the genus Bracon. The Chrysidæ are well represented, and the coloured wings of these species of Bracon are significant of the great heat of their dwelling-place.

The Insects were found in the following localities:—

Egypt.

Cairo. Shoobra. Heliopolis. Red Mountain. Geezech Pyramids.

Africa, near the Red Sea.

Berenice. Souakin. Hor Tamanib. Massowah. Sheykh Berout. Akeek (Island). Harkeko. Dahleck (Island). Rafla (Annesley

Bay). Tajura (Straits of Bab el Mandeb). Akceko (Arab village).
Zayla (Indian Ocean).

Arabia.

Wâdy Gennèh. Wâdy Amara. Wâdy Sidri. Wells of Moses.
Pharaoh's Baths. Wâdy Ferran. Wâdy Nash. Wâdy-es-Shekh.
Tôr. Wâdy Gharandel. Wâdy Hebran. Plain of Ramleh.
Gebel Musa, Mount Sinai. Gardens round Mount Sinai. Sandy
Plain, Mount Sinai. Convent garden, Mount Sinai. Wâdy Atall.

LIST

EGYPTIAN AND ARABIAN HYMENOPTERA.

Fam. EVANIIDÆ.
Genus FOENUS, *Fabr.*

1. FOENUS JACULATOR? Ichneumon jaculator, *Linn. Syst. Nat.* ii. 937.

The specimen here recorded is mutilated.

Tajura.

Inhabits Europe.

Fam. ICHNEUMONIDÆ.
Subfam. CRYPTIDES.
Genus CRYPTUS, *Fabr.*

2. CRYPTUS LEUCOPYGUS. *Female.*—Red, stout, very finely punctured. Head transverse, a little broader than the thorax. Antennæ black, much shorter than the body, with a broad white band beyond the middle. Thorax with two nearly parallel furrows; scutellum small, nearly smooth. Metathorax scabrous, large. Abdomen piceous, as long as the thorax; petiole widening from the base to the hind border: tip white. Oviduct black, about half the length of the abdomen. Legs piceous; hind tibiæ white at the base. Wings cinereous; veins black. Fore wings with a black stigma; intermediate cubital areolet minute, tetragonal. Length of the body 4 lines.

Allied to C. minutorius, *Fabr.*, and to C. bimaculatus, *Grav.*; distinguished from the former by the red head, from the latter by the unspotted wings.

Cairo.

B

Subfam. OPHIONIDES.

Genus ENICOSPILUS, *Steph.*

3. ENICOSPILUS RAMIDULUS.   Ichneumon ramidulus, *Linn. Fann. Suec.* 1029.

Harkeko.

Inhabits Europe.

Genus CAMPOPLEX, *Grav.*

4. CAMPOPLEX TARSALIS.   *Female.*—Black, slender, extremely minutely punctured.   Head about the eyes, two spots on the front and a spot on each side of the mouth, pale yellow.   Palpi pale yellow.   Antennæ stout, shorter than the body.   Thorax with a yellow stripe on each side; scutellum with a yellow spot on each side.   Meso-thorax with one yellow spot.   Metathorax finely scabrous; disk depressed, with a slight keel; a yellow spot on each side.   Abdomen slightly compressed, dark red beneath; segments successively decreasing in length.   Legs red; hind tibiæ blackish towards the tips; hind tarsi blackish, pale testaceous towards the tips.   Wings cinereous, veins black.   Fore wings with a black stigma; intermediate cubital areolet pentagonal.   Length of the body 5½ lines.

Cairo.

5. CAMPOPLEX POSTICUS.   *Male.*—Black, slender, extremely finely punctured.   Palpi yellow.   Antennæ rather stout.   Thorax dull, with two slight furrows, which converge slightly hindward.   Metathorax finely scabrous.   Abdomen slightly compressed, slightly deepening from the base to the tip, nearly twice the length of the thorax.   Legs red; hind tarsi blackish; first joint red towards the base; second, third and fourth joints pale yellow at the base.   Wings cinereous; veins black; root yellowish white.   Fore wings with a piceous stigma; intermedial cubital areolet tetragonal, small.   Length of the body 4 lines.

Cairo.

Subfam. TRYPHONIDES.

Genus BASSUS, *Fabr.*

6. BASSUS LÆTATORIUS.   Ichneumon lætatorius, *Fabr. Ent. Syst.* ii. 147.

Cairo.

Inhabits Europe.

Subfam. PIMPLIDES.

Genus PIMPLA, *Fabr.*

7. PIMPLA INSTIGATOR. Ichneumon instigator, *Fabr. Ent. Syst.*
ii. 164.

Cairo.

Inhabits Europe.

Genus XORIDES, *Latr.*

8. XORIDES ÆGYPTIUS. *Female.* — Black, slightly shining,
minutely punctured. Head smooth. Antennæ rather stout.
Thorax with two distinct furrows which slightly diverge in front.
Abdomen deep red, sometimes black towards the tip. Oviduct
nearly six lines in length and nearly as long as the body. Legs red.
Wings slightly smoky; veins black. Fore wings with a whitish
point at the base of the stigma; intermediate cubital areolet tetragonal, its outer side very short. Length of the body 5½ lines.

Cairo.

Fam. BRACONIDÆ.

Genus BRACON, *Fabr.*

9. BRACON FASTIDIATOR. Ichneumon fastidiator, *Fabr. Ent. Syst.*
ii. 156.

Cairo. Harkeko.

10. BRACON SCULPTURATUS. *Female.*—Black. Head and thorax
red, shining. Eyes and antennæ black; the latter stout, nearly as
long as the body. Metathorax black, shining, Abdomen fusiform,
very much longer than the thorax, elaborately and beautifully
sculptured; each segment with a striated triangular compartment and
a transverse ridge; under side with three pale yellow patches. Oviduct red, black towards the tip, longer than the body. Fore legs red,
their femora black towards the base. Wings black. Fore wings
with a luteous stigma, whence a white streak extends to a white dot
in the disk; an exterior white transverse streak. Length of the body
7½ lines.

The black abdomen of this species distinguishes it from B. fastidiator.

Hor Tamanib.

11. BRACON MELANARIUS. *Male and female.*—Black. Head and
thorax smooth, shining, pilose. Antennæ stout, a little longer than
the body. Abdomen red, sculptured much like that of B. sculpturatus.

Sheaths of the oviduct black, rather more than half the length of the abdomen. Wings black. Fore wings with an ochraceous stigma. Length of the body 5—6 lines.

Hor Tamanib. Massowah.

12. BRACON ARDENS. *Female.*—Bright red. Head and thorax smooth, shining. Antennæ stout, black, almost as long as the body. Abdomen finely scabrous. Oviduct black, rather more than half the length of the abdomen. Wings lurid. Fore wings with a blackish band joining the inner end of the stigma; surface from the outer end of the stigma to the tip blackish; veins red, black in the blackish part; stigma bright red. Hind wings with the exterior half obliquely blackish. Length of the body 4½ lines.

Tajura.

13. BRACON CONCOLOR. *Female.* — Red. Head and thorax smooth, shining, pilose. Head black above. Antennæ black, a little longer than the body. Abdomen finely striated. Sheaths of the oviduct black, about twice the length of the abdomen. Wings blackish. Fore wings with a pale luteous band at a little beyond the middle, and with an exterior pale cinereous spot in the disk. Hind wings with two pale luteous spots, the inner one in the disk, the outer one joining the costa. Length of the body 5 lines.

Hor Tamanib.

14. BRACON DETERMINATUS. *Male.*—Bright red. Head and thorax smooth, shining. Head and sides of the prothorax black. Antennæ black, stout, a little longer than the body. Abdomen very finely scabrous. Wings lurid, with a black middle band, black also between the stigma and the tips; veins red, black in the black part. Length of the body 4—4½ lines.

Hor Tamanib.
Raīla, Annesley Bay.

15. BRACON CONGRUUS. *Male.*—Red. Head and thorax smooth, shining. Head black about the ocelli. Antennæ black, stout, as long as the body. Abdomen finely striated. Wings black, lurid and with red veins along the costa towards the base. Fore wings with an irregular whitish band adjoining the stigma, which is red. Length of the body 6 lines.

Cairo.

16. BRACON SIGNIFER. *Male and female.*—Red, smooth, shining. Antennæ black, as long as the body. Sheaths of the oviduct black, rather longer than the abdomen. Wings black. Fore wings with a white band behind the stigma, which is red. Hind wings cinereous towards the base. Length of the body 3¼ lines.
Tajura.

17. BRACON INDECISUS. *Female.*—Red, smooth, shining. Front minutely punctured. Antennæ black, stout. Abdomen minutely punctured, shorter than the thorax. Oviduct and its sheaths black. Wings brown, with a cinereous middle band; stigma luteous, black towards the tip, and there contiguous to a dark brown spot. Length of the body 3½ lines.
Cairo.

18. BRACON XANTHOMELAS. *Female.*—Black, shining. Mouth rostriform. Antennæ stout. Thorax ochraceous-red; scutellum and metathorax black. Abdomen ochraceous-red, with a short petiole. Sheaths of the oviduct black, as long as the abdomen. Wings blackish; veins black. Length of the body 2 lines.
Nearly allied to B. urinator.
Cairo.

19. BRACON SPILOGASTER. *Male.*—Black, shining. Mouth rostriform. Antennæ stout. Abdomen luteous, fusiform, with a very short petiole and with a row of black dorsal spots, the three posterior spots larger and more determinate than the others. Wings cinereous; veins and stigma black. Length of the body 1¼ line.
Cairo.

Genus PHYLAX, *Wesm.*

20. PHYLAX? NIGRICORNIS. *Female.* — Testaceous, minutely pubescent. Eyes black, subelliptical, rather large and prominent. Antennæ black, stout. Abdomen compressed, subsessile, deepening from the base to the tip. Sheaths of the oviduct piceous, about one-sixth of the length of the abdomen. Hind tarsi and tips of hind tibiæ blackish. Wings cinereous, minutely pubescent; veins black, testaceous at the base. Length of the body 3½ lines.
Harkeko.

Genus MICROGASTER, *Latr.*

21. MICROGASTER FALCATUS, *Nees, Hym. Ich. aff.* ii. 175.
Cairo.
Inhabits North Europe.

Fam. CYNIPIDÆ.
Genus FIGITES, *Latr.*

22. FIGITES INAPERTUS. *Male.*—Black, smooth, shining. Antennæ submoniliform, a little longer than the body. Abdomen compressed, subsessile. Legs red; femora black, red at the tips. Wings pellucid; veins pale testaceous. Length of the body ¾ line.
Cairo.

Fam. CHALCIDIDÆ.
Genus CHALCIS, *Fabr.*

23. CHALCIS INSOLITA. *Male and female.*—Red. Head and thorax roughly punctured. Head black, with silvery cinereous hairs about the eyes. Antennæ black, stout, red at the base. Thorax thinly covered with hoary hairs. Abdomen black, smooth, shining, elliptical in the male, elongate-oval and acute in the female. Hind legs black, excepting the coxæ. Wings cinereous. Fore wings tinged with brown near the stigma, which is black. Length of the body 2½—3 lines.

Smaller than C. rubens (*Klug, Symb. Phys.* pl. 37, f. 7); distinguished also from that species by the black head and black abdomen and the more slender antennæ.
Tajura.

Fam. CHRYSIDÆ.
Genus PARNOPES, *Latr.*

24. PARNOPES APICALIS. Golden green. Head and thorax scabrous, with whitish pubescence. Front slightly excavated; mouth black, rostriform. Antennæ black; first joint green. Hind part of the thorax with a purple tinge. Abdomen purple towards the tip; apical segment with white pubescence, and with a dentate and slightly retuse border; under side black. Legs green; knees, tarsi and tips of tibiæ tawny. Wings cinereous; veins black. Fore wings brownish on more than one-third of the apical part. Length of the body 4½ lines.
Tajura.

Genus EUCHRÆUS, *Latr.*

25. EUCHRÆUS PALLISPINOSUS. Golden green, with short whitish pubescence. Head·purple on each side between the eyes and the hind border. Antennæ black; first joint purple. Thorax purple, with a broad golden green stripe, of which the fore part has a cupreous disk. Abdomen bluish green, with a broad transverse subapical

furrow; tip green, with eleven pale testaceous spines. Legs bluish green; knees, tarsi and tips of tibiæ pale testaceous. Wings pellucid; veins black. Length of the body 4 lines.
Cairo.

Genus STILBUM, *Spinola.*

26. STILBUM SPLENDIDUM. Chrysis splendida, *Fabr. Ent. Syst.* ii. 238.
Cairo. Tajura. Wâdy Ferran.
Inhabits South Europe.

Genus CHRYSIS, *Linn.*

27. CHRYSIS AUSTRIACA, *Fabr. Syst. Piez.* 173.
Cairo.

28. CHRYSIS VARICORNIS, *Spin.*
Cairo.

29. CHRYSIS ALTERNANS, *Klug.*
Wâdy Genneh.

30. CHRYSIS SINAICA. *Female.*—Purple, thickly punctured. Head bright cupreous between the eyes; face black. Eyes piceous. Antennæ black; first joint bluish green. Scutum cupreous, with two parallel furrows. Metathorax with a stout spine on each side. Abdomen with three large segments; third segment green at the base, near which there is a slight transverse depression; tip quadridentate. Tarsi black. Wings cinereous; veins black. Length of the body 5½ lines.
Mount Sinai.

31. CHRYSIS SEMINIGRA. *Female.*—Black, thickly punctured, with cinereous tomentum. Head purple in front. Antennæ black; first joint purple. Scutum with two slight parallel furrows. Metathorax with a stout spine on each side of the hind border. Pectus purple. Abdomen golden green, with a very slight keel and with three large segments; third segment deep black at the base; tip quadridentate; under side black. Legs purple; tarsi black. Wings cinereous; veins black. Fore wings smoky brown, except along the exterior border. Length of the body 5 lines.
Wâdy Ferran.

32. CHRYSIS MULTICOLOR. *Male.*—Purple, thickly punctured, with cinereous pubescence. Head green in front. Antennæ black; first joint green. Scutum black between the two distinct parallel furrows. Metathorax with a stout spine on each side of the hind border. Abdomen cupreous, with a slight keel and with three large segments; first segment, and the hind borders of the second and third segments, golden green; hind border of the third segment golden green; under side purple, partly blue and green. Legs bluish green; tarsi black. Wings cinereous; veins black. Length of the body 5 lines.

Wâdy Ferran.

Female.— Bluish green. Head purple about the hind border; triangle of the ocelli black. Antennæ black; first joint bluish green. Scutum purple between the two parallel furrows. Abdomen cupreous; first segment golden green; under side green. Legs green; tarsi tawny. *Var. β.*—Golden green. Scutum black between the two parallel furrows. Abdomen as in *Var. α,* except the disk of the first segment, which is cupreous. Tarsi piceous. Length of the body 3—3½ lines.

Wâdy Ferran.

33. CHRYSIS ELECTA. *Male.* — Golden green, with cinereous pubescence, thickly punctured. Head with silvery pubescence on each side in front. Scutum purple between the two parallel furrows. Metathorax with a stout tooth on each side of the hind border. Abdomen with three large segments and with an extremely slight keel; second and third segments blue about the fore border; hind border of the third with four teeth. Legs green; tarsi piceous, pale testaceous at the base. Wings pellucid; veins black. *Var. β.*— Tarsi wholly piceous. Length of the body 4—4½ lines.

Wâdy Ferran.

34. CHRYSIS COMMUNIS. *Male and female.*—Green, with cinereous pubescence, thickly punctured. Antennæ black; first joint green. Scutum bluish green; the two parallel furrows well defined. Hind border of the metathorax with a stout spine on each side. Abdomen with three large segments and with an extremely slight keel; hind border of the third segment quadridentate. Tarsi black. Wings cinereous; veins black. *Var. β.*—Bluish green. *Var. γ.*—Purplish blue. Length of the body 2½—3 lines.

Wâdy Ferran. Tajura.

Genus PYRIA, *Lep. et Serv.*

35. PYRIA SMARAGDULA, *Lep. et Serv. Enc. x. 494*
Wády Ferran.
Inhabits Egypt and Senegal.

Gen. HEDYCHRUM, *Latr.*

36. HEDYCHRUM STILBOIDES. *Male and female.*—Blue, punctured, with cinereous pubescence, partly purple and green. Antennæ black; first joint blue. Eyes piceous. Metathorax twice as broad as long, with a stout spine on each hind angle. Abdomen with three large segments, bidentate at the tip. Tarsi piceous. Wings blackish; veins black. Length of the body 3½—4½ lines.
Cairo.

ACULEATA.

Tribe HETEROGYNA.

Fam. FORMICIDÆ.

Genus FORMICA, *Linn.*

37. FORMICA ÆQUALIS. *Female.*—Luteous, thinly setose. Head somewhat broader than the thorax; clypeus with long setæ. Eyes black, elliptical. Mandibles very acute, black towards the tips, crossing each other transversely. Palpi long, slender. Petiole of the abdomen with a stout lanceolate upright spine. Abdomen elliptical, rather broader than the thorax. Legs slender, rather long. Length of the body 4 lines.
Wády Nash.

Genus CAMPONOTUS, *Mayr.*

38. CAMPONOTUS LIGNIPERDA. Formica ligniperda, *Latr. Hist. Nat. Fourm.* 88.
Wády Ferran. Mount Sinai.
Inhabits Europe.

39. CAMPONOTUS SERICEUS. Formica sericea, *Fabr. Ent. Syst. Supp.* 279.
Mount Sinai.
Inhabits West Africa.

c

40. CAMPONOTUS PHÆOGASTER. *Neuter.*—Tawny, smooth, with silvery cinereous tomentum. Head very much broader than the thorax. Eyes small, black, elliptical, slightly prominent. Mandibles long, slender, acute, black towards the tips. Antennæ piceous towards the tips. Prothorax and mesothorax conical, the former much broader than the latter. Metathorax subquadrate, a little broader than the mesothorax. Petiole forming a high transverse ridge. Abdomen piceous, elliptical, much shorter than the thorax. Legs long, slender. Length of the body 4 lines.

Wâdy Ferran.

Genus CATAGLYPHIS, *Foerst.*

41. CATAGLYPHIS VIATICA. Formica viatica, *Fabr. Ent. Syst.* ii. 356.

Wâdy Amara. Mount Sinai.

Inhabits Europe.

42. CATAGLYPHIS BICOLORIPES. *Female.*—Red. Head minutely punctured, much broader than the thorax. Eyes black, small, prominent, nearly round. Ocelli approximate, forming a triangle on the vertex. Palpi long, slender. Antennæ blackish, slender, more than half the length of the body; first and second joints red. Petiole nodose or gibbous above. Abdomen black. Legs tawny, long, slender; coxæ, trochanters and femora black, tips of the latter tawny. Length of the body 2¾ lines.

Cairo.

Fam. ATTIDÆ.
Genus APHŒNOGASTER, *Mayr.*

43. APHŒNOGASTER STRUCTOR. Formica structor, *Latr. Hist· Nat. Fourm.* 236.

Cairo. Wells of Moses.

Inhabits Europe.

44. APHŒNOGASTER DEBILIS. *Neuter.*—Black, slender. Head large, nearly twice the breadth of the thorax. Eyes small, flat, nearly round. Antennæ subclavate, a little more than half the length of the body. Thorax dark red. Petiole binodose, or forming two humps above, the anterior hump higher than the posterior one. Abdomen nearly oval, slightly convex, nearly twice the breadth of the thorax, truncate at the tip; first segment occupying nearly the whole surface. Legs long, very slender. Length of the body

1½ line. *Female.*—Antennæ and legs testaceous, the former black at the base. Wings whitish; veins pale testaceous. Length of the body 2 lines.

Pharaoh's Baths.

45. APHŒNOGASTER PALLESCENS. ' *Female.* — Black, shining. Head tawny about the mouth. Eyes flat. Antennæ, abdomen and legs testaceous. Antennæ short, subclavate. Petiole with two humps, like those of A. debilis. Abdomen more than twice the breadth and nearly twice the length of the thorax. Wings whitish, long; veins testaceous. Length of the body 2¼ lines.

Cairo.

Fam. MUTILLIDÆ, *Smith.*

Genus MUTILLA, *Linn.*

46. MUTILLA AUREIVENTRIS. *Male.*—Black. Head, thorax and legs clothed with silvery hairs. Head and thorax thickly punctured. Prothorax and mesothorax red. Abdomen luteous, clothed with gilded hairs. Wings bluish black. Length of the body 8 lines.

Hor Tamanib.

Genus APTEROGYNA, *Latr.*

47. APTEROGYNA OLIVIERI, *Latr. Gen. Ins.* iv. 122.

Pharaoh's Baths.

Fam. SCOLIIDÆ.

Genus MYZINE, *Latr.*

48. MYZINE GUERINEI, *Lucas, Expl. Sci. Alg.* iii. 284, pl. 15, f. 6.

Cairo.

49. MYZINE FLAVICOLLIS. *Female.*—Black, clothed with short white hairs. Head and thorax thickly and minutely punctured. Head yellow about the mouth. Antennæ red beneath. Prothorax, a spot on the scutum, disk of the scutellum, tegulæ of the fore wings and a large spot on each of the pleuræ yellow. Metathorax with two yellow dots on the base. Abdomen with seven entire yellow bands above and beneath, the seventh apical. Aculeus curved upward. Legs yellow; femora above with an abbreviated black stripe. Wings pellucid, veins black, pale yellow at the base. Fore wings with a tawny stigma. Length of the body 5 lines.

It differs from M. Oraniensis by the abdomen being not red at the base.

Wâdy Gennêh.

50. Myzine punctifascia. *Female.*—Black, clothed with whitish hairs. Head and prothorax thickly punctured. Abdomen thinly punctured, with five narrow yellow bands, of which the fore border is notched on each side; first and second bands dilated and including a black dot on each side; under side with two rows of transversely elongated yellow spots. Aculeus curved upward. Legs yellow; femora black; fore femora yellow towards the tips; tibiæ black beneath; tarsi with a brown band on the end of each joint and with brown tips. Wings pellucid; veins black. Fore wings with a black stigma. Length of the body 5 lines.

The red legs of M. nitida distinguish it from this species.

Wâdy Gennèh.

Genus DISCOLIA, *Sauss.*

De Saussure's monograph of the Scoliadæ includes some new subgenera, and they are here recorded.

Genus Liacos, *Guer.*   Second discoidal areolet petiolated in the second cubital areolet.

Subgenus 1. Triliacos.   Three entire cubital areolets.

Subgenus 2. Diliacos.   Two complete cubital areolets.

Genus Scolia, *Fabr.*   One recurrent vein.

Subgenus 1. Triscolia.   Three complete cubital areolets.

Subgenus 2. Discolia.   Two complete cubital areolets.

Genus Elis, *Fabr.*   Two recurrent veins.

Subgenus 1. Trielis.   Three complete cubital areolets.

Subgenus 2. Dielis.   Two complete cubital areolets.

51. Discolia lateralis.   Scolia lateralis, *Klug, Symb. Phys.* pl. 26, f. 3, 4.

Akeek.   Harkeko.

Inhabits Cyprus.

52. Discolia maura.   Scolia maura, *Fabr. Ent. Syst.* ii. 233.

Wâdy Hebran.   Mount Sinai.

Inhabits South Europe and North Africa.

53. Discolia Hottentotta, *Sauss. Cat. Scol.* 89.

Wâdy Gennèb.

Inhabits South Africa.

54. Discolia mendica. Scolia mendica, *Klug, Symb. Phys.* pl. 26, f. 15.

Cairo.

55. Discolia insubrica. Scolia insubrica, *Rossi, Faun. Etrus.* ii. 72.

Massowah.

Inhabits South Europe.

56. Discolia abyssinica, *Sauss. Cat. Scol.* 87.

Massowah. Harkeko.

Inhabits Cyprus.

57. Discolia luteicornis. *Male.*— Bluish black, punctured, clothed with whitish hairs. Head with a tawny line along the hind side of each eye. Mandibles ferruginous. Antennæ luteous; first and second joints black. Abdomen with three very slender apical spines. Wings black, iridescent. Length of the body 6—6½ lines.

The black head distinguishes it from the variety of D. erratica with red antennæ. It differs from D. instabilis in its smaller size, in its paler antennæ and in its wholly black colour.

Massowah. Harkeko. Akeek.

58. Discolia atratula. *Male.*—Deep black, punctured, with short cinereous hairs. Abdomen bluish black, with short black hairs; tip with three slender spines. Wings purplish black. Length of the body 5—5½ lines.

Smaller than D. læviceps, from which it is distinguished also by the cinereous hairs on the front.

Wâdy Gennêh.

Genus DIELIS, *Sauss.*

59. Dielis collaris. Tiphia collaris, *Fabr. Ent. Syst.* ii. 227.

Mount Sinai.

60. Dielis auricollis. Campsomeris aureicollis, *St. Farg. Hym.* iii. 499.

Hor Tamanib. Massowah. Akeek. Rafla.

De Saussure observes that D. auricollis is a synonym of Tiphia thoracica, *Fabr.*; the latter inhabits Hindostan, China and Java.

61. DIELIS AUREOLA. Elis aureola, *Klug, Symb. Phys.* pl. 27, f. 11.

Massowah. Harkeko. Dahleck. Tajura.

Inhabits Natal.

62. DIELIS LONGISPINA. *Male.*—Black, clothed with rather long white hairs. Face yellow, with a black spot near the mouth. Thorax with a yellow fore border and with three yellow spots on the scutellum; of these the third or hinder one is transverse. Abdomen with five yellow bands ; of these, excepting the first, the fore borders are notched on each side. Legs yellow ; femora black, with a yellow stripe and with yellow tips; tibiæ striped with black; tips of the tarsal joints tawny. Wings pellucid ; veins brown. Fore wings brownish towards the tips. Length of the body 5½ lines.

The bands of the abdomen are much narrower than those of D. aureola, and the spines at the tip are longer. The partly yellow femora distinguish it from D. fasciatella.

Wâdy Gennêh.

Genus TRISCOLIA, *Sauss.*

63. TRISCOLIA BIDENS. Sphex bidens, *Linn. Syst. Nat.* i. 943.

Cairo.

Inhabits South Europe.

Genus TRIELIS, *Sauss.*

64. TRIELIS ALIENA. Scolia aliena, *Klug, Symb. Phys.* pl. 27, f. 3.

Massowah. Akeek. Harkeko.

Genus TIPHIA, *Fabr.*

65. TIPHIA LATIPES. *Female.*—Black, clothed with cinereous hairs. Head and thorax largely punctured. Head broader than the thorax. Mandibles ferruginous towards the tips. Antennæ thick. Abdomen fusiform, finely punctured. Legs rather short; four posterior femora and tibiæ dilated ; four posterior tibiæ dentate on the outer side. Wings black. Fore wings as long as the abdomen. Length of the body 5 lines.

The wings are darker than those of T. compressa and of T. nitida.

Massowah.

Fam. POMPILIDÆ.
Genus POMPILUS, *Fabr.*

66. POMPILUS FUSCUS. Sphex fusca, *Linn. Syst. Nat.* i. 944
Heliopolis.
Inhabits Europe.

67. POMPILUS CLYPEATUS, *Klug, Symb. Phys.* pl. 39, f. 14.
Cairo. Hor Tamanib. Tôr.

68. POMPILUS GLABRATUS, *Klug, Symb. Phys.* pl. 38, f. 1.
Akeek.
Inhabits West Africa.

69. POMPILUS PLUMBEUS. Sphex plumbea, *Fabr. Ent. Syst. Supp.*
ii. 220.
Hor Tamanib.
Inhabits Europe and North Africa.

70. POMPILUS INNITENS. *Female.*—Deep black, not shining,
clothed with black hairs. Antennæ curved. Head and thorax
finely punctured. Metathorax with a distinct keel. Abdomen fusiform, extremely finely punctured. Wings blackish, veins black.
Length of the body 9 lines.

The tarsi are longer than those of P. niger and have black spines.
The more slender antennæ distinguish it from P. vindicatus. It
more closely resembles the American P. nebulosus, but the latter has
not a furrow on the metathorax.
Wâdy Gennêh.

71. POMPILUS MELANOPHILUS. *Female.* — Deep black, slightly
shining, rather stouter than P. innitens. Head and thorax finely
punctured. Antennæ curved. Metathorax scabrous, not keeled.
Abdomen elliptical, extremely finely punctured. Wings black; veins
deep black. Length of the body 7¼ lines.
Pharaoh's Baths.

Genus PRIOCNEMIS, *Schiodte.*

72. PRIOCNEMIS BRUNNEUS. Pompilus brunneus, *Klug, Symb.
Phys.* pl. 38, f. 2.
Cairo.

Genus AGENIA, *Schiodte.*.

73. AGENIA BIZONATA. *Female.*—Black, dull. Mandibles red. Antennæ stout, curved. Metathorax extremely finely punctured, not keeled. Petiole short. Wings cinereous; veins and stigma black. Fore wings with two blackish bands, the first in the middle, the second nearer the tip. Length of the body 4 lines.

The wings are darker and their bands are more blackish and more defined than those of A. maculipes.

Cairo.

74. AGENIA TERMINALIS. *Male.* — Black, dull, with cinereous tomentum. Mandibles red towards the tips. Metathorax with silvery pile, extremely finely punctured, not keeled. Abdomen with a white apical spot above. Legs red; fore femora blackish above; tarsi blackish. Wings cinereous, blackish at the tips; veins and stigma black. Length of the body 3½ lines.

The dark tips of the wings distinguish this species from A. mutabilis.

Wâdy Ferran.

75. AGENIA DECORA. *Male.*—Black, with cinereous tomentum. Clypeus with a whitish border, which is notched in the middle. Mandibles whitish, with piceous tips. Palpi testaceous. Antennæ red. Petiole very short. Abdomen with two interrupted white bands and with a white apical spot. Legs red; hind tibiæ with a white stripe above; spurs white; four posterior tarsi black, white like the fore tarsi towards the base. Wings cinereous, with blackish tips and a blackish stigma; veins black. Length of the body 3¼ lines.

Harkeko.

76. AGENIA TRISTIS. *Male.*—Black, dull. Petiole very short. Wings cinereous, blackish towards the tips; veins and stigma black. Length of the body 3—3½ lines.

Mount Sinai.

Genus EVAGETHIS, *St. Farg.*

77. EVAGETHIS BICOLORIFER. *Male.* — Black, with cinereous tomentum. Mandibles red; tips black. Metathorax subquadrate, with a longitudinal furrow. Petiole extremely short. Abdomen red, subfusiform. Four posterior femora and tibiæ red; fore femora red

towards the tips; fore tibiæ red towards the base; fore tarsi armed
with five long slender spines on the outer side. Wings cinereous,
blackish towards the tips; veins and stigma black. Length of the
body 4 lines.
Hor Tamanib.

Genus SALIUS, *Fabr.*

78. SALIUS BICOLOR, *Fabr. Syst. Piez.* 124.
Rafla.
Inhabits Albania and North Africa.

Genus FERREOLA, *St. Farg.*

79. FERREOLA DIVISA. *Male.*—Deep black. Head red; a black
spot on the vertex including the ocelli. Eyes black, fusiform.
Antennæ black; first and second joints red. Thorax red. Prothorax
with gilded tomentum. Fore border of the scutum of the mesothorax
with a blackish abbreviated band which emits two piceous longitudinal streaks; scutellum elongate-conical, slightly notched and
truncate at the tip. Metathorax black. Legs red; coxæ, trochanters,
four posterior femora at the base, four posterior tibiæ and four
posterior tarsi black. Wings bluish black; veins black; tegulæ red.
Length of the body 6 lines.

The red legs distinguish it from F. dimidiata, from F. Schiodtii,
and from F. bicolor.
Hor Tamanib. Tajura.

80. FERREOLA CARBONARIA. *Male.*—Deep black. Head dark
red about the eyes and along the hind border. Eyes piceous, fusiform.
Prothorax with a large dark red patch on each side. Wings bluish
black; veins black. Length of the body 6 lines.
Mount Sinai.

Genus AMMOPHILA, *Kirby.*

81. AMMOPHILA ARGENTEA, *Brullé, Hist. Nat. Il. Canar.* iii. 65.
Hor Tamanib. Harkeko. Mount Sinai.
Inhabits the Canary Isles.

82. AMMOPHILA NASUTA, *St. Farg. Hym.* iii. 380.
Massowah. Akeek. Rafla. Tajura.
Inhabits Portugal.

83. AMMOPHILA RUBRIPES, *Spinola, Ann. Soc. Ent. Fr.* vii. 465.
Massowah. Wâdy Gennèh.

D

84. AMMOPHILA FERRUGINEIPES, *St. Farg. Hym.* iii. 383.

Hor Tamanib. Tajura.

Inhabits West Africa and South Africa.

85. AMMOPHILA EBERRINA, *Spinola, Ann. Soc. Ent. Fr.* vii. 464.

Mount Sinai.

Inhabits Egypt.

86. AMMOPHILA STRENUA. *Female.*—Black, stout, clothed with black hairs. Head rather broader than the thorax. Metathorax with cinereous hairs about its hind border. Petiole black, slender, subclavate, a little shorter than the metathorax. Abdomen red, fusiform, black towards the tip. Wings cinereous, blackish towards the tips; veins black. Length of the body 8½—9½ lines.

The black front distinguishes it from A. argentata; it is more slender than A. viatica.

Cairo. Wâdy Gennêh. Wâdy Ferran.

87. AMMOPHILA FILATA. *Male.*—Black, very slender. Head, thorax and petiole covered with silvery tomentum. Mandibles red, black towards the tips. Mouth mostly red. Antennæ red at the base. Petiole very slender, much longer than the metathorax, and a little longer than the following segment of the abdomen. Abdomen red, clavate, slightly compressed, black towards the tip, longer than the thorax, inclusive of the petiole. Legs red; hind femora towards the base and coxæ black. Wings pellucid, short; veins black, red towards the base. Length of the body 9 lines.

The more silvery colour, the black petiole and the wholly red hind legs distinguish it from A. nasuta.

Wâdy Gennêh.

88. AMMOPHILA NIGRITARIA. *Male and female.*—Black, slender. Head and thorax with cinereous hairs. Front of the head with silvery tomentum. Petiole very slender, much longer than the metathorax. Abdomen clavate, about as long as the thorax and the petiole together; second segment with a red line on each side; third with a red patch on each side. Wings pellucid; veins black. Fore wings as long as the thorax and the petiole together. Length of the body 9—10 lines.

·Closely allied to the Albanian A. dives; the latter has no red marks on the abdomen.

Tajura.

89. Ammophila areolata. *Male.*—Black. Head and thorax
minutely punctured, clothed with black hairs. Head rather broader
than the thorax. Metathorax much developed. Petiole slender,
about half the length of the metathorax. Abdomen elongate-elliptical,
smooth, shining, as long as the thorax, exclusive of the petiole.
Wings smoky; veins black. Fore wings longer than the head,
thorax and petiole together; some of the areolets with paler disks.
Length of the body 6½ lines.
Pharaoh's Baths.

Genus PELOPŒUS, *Latr.*

90. Pelopœus spirifex. Sphex spirifex, *Linn. Syst. Nat.* i. 942.
Cairo.
Inhabits South Europe.

91. Pelopœus violaceus. Sphex violacea, *Fabr. Ent. Syst.* ii
201.
Massowah. Harkeko. Wâdy Gennèh. Wâdy Ferran. Mount
Sinai.
Inhabits Asia Minor.

Genus SPHEX, *Linn.*

92. Sphex argentata, *Dahlb. Hym. Eur.* i. 25
Dahleck. Tajura.
Inhabits Europe, Africa, Asia and North America.

93. Sphex argentifera. *Male.*—Black, stout, with cinereous
tomentum. Head a little broader than the thorax, with silvery
hairs behind; front with silvery tomentum. Metathorax with silvery
tomentum and thickly clothed with silvery hairs. Petiole slender,
silvery, not more than half the length of the metathorax. Abdomen
elliptical, as long as the thorax, exclusive of the petiole; first and
second segments especially tomentose. Wings cinereous; veins
black. Fore wings with blackish tips, much shorter than the body.
Length of the body 10 lines.
    Closely allied to A. argentata, but the metathorax is more densely
clothed with white hairs.
Akeek.

Genus HARPACTOPUS, *Smith.*

94. HARPACTOPUS CRUDELIS, *Smith, Cat. Hym.* iv. 264, pl. 6, f. 4. Akeek. Harkeko. Wâdy Hebran. Inhabits Madras.

Genus PARASPHEX, *Smith.*

95. PARASPHEX FERVENS. Sphex fervens, *Fabr. Ent. Syst.* ii. 200. Harkeko. Rafla. Tajura. Tôr. Inhabits West Africa, South Africa and Hindostan.

Genus CHLORION, *Latr.*

96. CHLORION MELANOSOMA, *Smith, Cat. Hym.* iv. 238. Wâdy Nash. Wâdy Ghârandel. Inhabits Hindostan.

97. CHLORION BICOLOR. *Female.*—Black, shining, smooth, stout, with some black hairs. Head red, broader than the thorax. Mandibles black. Antennæ red. Prothorax red above. Metathorax very large, transversely and very finely striated. Petiole less than one-fourth of the length of the metathorax. Abdomen deep blue, elongate-oval, a little shorter than the thorax, exclusive of the petiole. Wings black, with blue and purple reflections, much shorter than the body. Length of the body 10½ lines.

Nearly allied to an undescribed species from Beluchistan in the British Museum.

Wâdy Ghârandel.

Fam. LARRIDÆ, *Steph.*

Genus LARRADA, *Smith.*

98. LARRADA ANATHEMA. Sphex anathema, *Rossi, Faun. Etr.* ii. 65. Cairo. Inhabits Europe and South Africa.

99. LARRADA HÆMORRHOIDALIS. Pompilus hæmorrhoidalis, *Fabr. Syst. Piez.* 198. . Cairo. Inhabits West Africa and Hindostan.

100. LARRADA NIGRITA. Tachytes nigrita, *St. Farg. Hym.* iii. 241.
Cairo. Wâdy Ferran. Mount Sinai.
Inhabits Madeira.

101. LARRADA ORANIENSIS. Tachytes Oraniensis, *St. Farg. Hym.* iii.
Tajura.

102. LARRADA NIGRICANS. *Male and female.* — Black, dull, with cinereous tomentum. Head a little broader than the thorax, silvery in front and behind. Petiole very short. Abdomen elliptical in the male, elongate-oval in the female. Wings blackish cinereous, blackish towards the tips; veins black. Length of the body 4—6 lines.

This species is a little smaller and has paler wings than L. nigrita, which it much resembles.

Cairo. Hor Tamanib.

103. LARRADA SUBFASCIATA. *Female.* — Black, with cinereous reflections. Head a little broader than the thorax, silvery behind and in front. Hind part of the metathorax striated, with a longitudinal furrow. Petiole very short. Abdomen elongate-oval, with four widely interrupted silvery bands. Tarsi dark red. Wings cinereous; veins black. Fore wings dark cinereous along the exterior border. Length of the body 6 lines.

Cairo.

104. LARRADA CONJUNGENS. *Female.* — Black, with . silvery cinereous tomentum. Head silvery in front and behind. Mandibles black. Petiole black, very short. Abdomen red, elongate-oval, longer than the thorax, with four silvery patches on each side: apical segment gilded above. Legs with silvery tomentum; hind femora dark red. Wings pellucid; veins tawny. Length of the body 7½ lines.

Dahleck.

105. LARRADA DIVERSA. *Female.*—Black, with silvery cinereous tomentum. Head silvery in front and behind. Mandibles black. Metathorax finely and transversely striated, with a slight longitudinal furrow. Petiole black, very short. Abdomen red, elongate-oval, a little longer than the thorax, with three silvery patches on each side; second segment with a black oblique streak on each side; fourth,

fifth and sixth segments black. Legs red ; coxæ and femora black, the latter with red tips ; fore tarsi with long slender black spines. Wings pellucid ; veins black, testaceous at the base. *Male ?*—Abdomen elongate-elliptical, not black towards the tip. Femora red, black above towards the base; fore tarsi without spines. Length of the body 5—6 lines.

Cairo.

Genus TACHYTES, *Panz.*

106. TACHYTES OBSOLETUS. Apis obsoleta, *Rossi, Faun. Etr.* i. 143.

Wâdy Gennêh. Wâdy Ferran.

Inhabits Europe.

107. TACHYTES PLAGIATUS. *Female.*—Red, very finely punctured. Head yellow and with silvery tomentum on each side between the eyes. Abdomen black, elongate-elliptical, subsessile, longer than the thorax. Petiole very thick, widening from the base to the tip, where there is a yellow band which is notched in the middle ; five posterior yellow bands, the fifth notched on each side ; tip yellow. Coxæ, trochanters, four posterior femora and hind tibiæ black ; fore femora black and with a yellow apical streak on the outer side ; four anterior tibiæ black on the outer side. Wings cinereous ; veins black. Fore wings with a blackish patch extending along the apical part of the costa and round the tip. Length of the body 4½ lines.

Hor Tamanib.

108. TACHYTES CONTRACTUS. *Male.*—Black. Head and thorax with silvery tomentum. Eyes cupreous, tessellated with black. Mouth, antennæ and legs red. Metathorax with a pale yellowish spine on each side at the base. Abdomen sessile, elliptical, longer than the thorax, with silvery cinereous tomentum at the base above and over most of the surface beneath ; a large yellow spot on each side near the base, and five yellow bands, the first interrupted ; some slender whitish curved tentacles on each side near the tip beneath. Legs red. Wings dark cinereous, blackish at the tips ; veins black. Length of the body 4 lines.

Dahleck.

109. TACHYTES LUGUBRIS. *Female.* — Black, with cinereous tomentum. Head broader than the thorax ; front silvery. Petiole rather stout, very short. Abdomen elliptical, a little longer than the thorax ; a pale gilded patch in the disk beneath. Fore tarsi with

reddish tips. Wings cinereous; veins black. Length of the body 4 lines.

Wâdy Ferran.

110. TACHYTES DECORATUS. *Female.* — Black, with cinereous tomentum. Head broader than the thorax; face and palpi yellowish white. Antennæ red, subclavate, first joint yellowish white. Fore border of the prothorax, tegulæ and an abbreviated band at the base of the metathorax yellowish white. Petiole red, with an abbreviated yellowish white band on its hind border; five posterior yellowish white bands, the first excavated on each side; under side with yellowish white lateral dots. Legs yellowish white; coxæ red; four anterior femora piceous, with a yellowish white streak beneath; hind femora black; four anterior tibiæ reddish beneath; hind tibiæ black beneath. Wings pellucid; veins black, reddish at the base. Length of the body 4 lines.

Wâdy Ferran.

111. TACHYTES CEPHALOTES. *Male.* — Black, with cinereous tomentum. Head broader than the thorax; front with gilded tomentum. Mouth and first joint of the antennæ red. Petiole black, very short. Abdomen red, elongate-oval, black towards the tip, not longer than the thorax. Legs red; coxæ black; femora with a black streak above. Wings pellucid; veins pale testaceous, black towards the tips; stigma black. Length of the body 3½ lines.

Harkeko.

112. TACHYTES ALBONOTATUS. *Female.*—Deep black, punctured. Front with silvery tomentum. Prothorax with four white dots on the fore border. Tegulæ white. Metathorax with a longitudinal furrow, its hind part perpendicular. Petiole sometimes red. Abdomen elliptical; first segment occupying nearly half the surface, its hind border white; a transverse white dot in the disk near the tip. Tibiæ with an elongated white spot above at the base; hind tarsi with three white dots above; four anterior tarsi whitish. Wings blackish; veins black. Length of the body 2½ lines.

Harkeko.

113. TACHYTES MUTILLOIDES. *Female.* — Black. Head red behind, pale gilded yellow about the eyes and in front. First joint of the antennæ white beneath. Thorax red. Pectus black. Abdomen fusiform, longer than the thorax; three white bands; first interrupted and abbreviated; second much excavated in the middle part of its fore side; third entire. Four anterior legs red; fore

femora interruptedly black above; middle femora wholly black above; middle tarsi black. Wings blackish; veins black. Length of the body 3 lines. Harkeko.

114. TACHYTES BREVIS. *Female.* — Black, broad, short, with cinereous tomentum. Head as broad as the thorax, with silvery tomentum about the sockets of the antennæ. Antennæ piceous, red beneath. Scutellum ferruginous. Petiole very short. Abdomen tawny, oval, longer than the thorax, with three black bands; first band basal; second near the base; third before the middle. Legs piceous, with whitish tomentum. Wings cinereous, dark cinereous about the tips; veins black. Length of the body 2½ lines. Tajura.

Fam. BEMBECIDÆ.

Genus BEMBEX, *Fabr.*

115. BEMBEX TREPANDA, *Dahlb. Hym. Eur.* i. 181. Tajura. Wâdy Gennèh. Wâdy Ferran. Inhabits Hindostan.

116. BEMBEX SULPHURESCENS, *Dahlb. Hym. Eur.* i. 180. Wâdy Ferran. Inhabits Hindostan.

117. BEMBEX REPANDA, *Latr. Gen. Ins.* iv. 98. Tajura. Inhabits South Europe.

118. BEMBEX OCULATA, *Jur. Hym.* 175, pl. 10. Hor Tamanib. Tajura. Mount Sinai. Inhabits South Europe.

119. BEMBEX OLIVACEA, *Fabr. Mant. Ins.* i. 285. Hor Tamanib. Tôr. Inhabits South Europe and North Africa.

Fam. NYSSONIDÆ.

Genus LARRA, *Klug.*

120. LARRA ZONATA, *Klug, Symb. Phys.* pl. 46, f. 2. Wâdy Ferran.

121. LARRA BIZONATA. Stizus bizonatus, *Spin. Ann. Soc. Ent. Fr.* vii. 473.

Akeek, Harkeko.

122. LARRA NUBILIPENNIS, *Smith, Cat. Hym.* iv. 347.
Wády Hebran.

123. LARRA TRIDENS. Crabro tridens, *Fabr. Ent. Syst.* ii. 298.
Tajura.
Inhabits South Europe and North Africa.

124. LARRA VESPOIDES. *Female.*—Piceous. Front of the head pale luteous, with silvery tomentum. Abdomen yellow; first segment black; second with a narrow black basal band, which on each side is dilated and connected with a narrow black band on the hind border of the same segment; third segment with a narrow black band on its hind border; fourth, fifth and sixth segments with a narrow red band on the hind border of each; the fore borders of these five bands are dentate in the middle; tip piceous, with a spine on each side; under side black, luteous and with blackish patches towards the tip. Wings cinereous; veins piceous. Fore wings brown along the costa towards the tips. Length of the body 14 lines.
Rafla.

125. LARRA SUBAPICALIS. *Female.*—Black, with whitish hairs. Orbits of the eyes, front and face yellow; a black dot on each side of the fore border of the front. Antennæ red, piceous above towards the tips; first joint yellow. Thorax very finely punctured; fore border, tubercles, a band on the scutellum with a much excavated fore border, and an interrupted band on the metathorax yellow. Abdomen with five yellow bands; first interrupted; second, third and fourth occasionally interrupted; under side also with variable bands. Legs yellow; femora towards the base and coxæ black. Wings slightly cinereous; veins tawny, piceous towards the tips. Fore wings with a brown subapical patch. Length of the body 6½—7 lines.

Nearly allied to L. ruficornis.
Wády Ferran.

126. LARRA LATIFASCIA. *Female.*—Black. Head with silvery tomentum, yellow and more brilliantly silvery between the insertion of the antennæ and the mouth. Antennæ yellow. Thorax bordered with testaceous in front and on each side and more broadly on the

E

hind border of the scutum; scutellum with a crescent-shaped testaceous band on its fore border. Metathorax with a very large yellow spot on each side. Pectus with a broad testaceous band. Abdomen with six yellow bands; first band deeply excavated in the middle of the fore border; the five following more or less undulating along the fore border; tip reddish; under side wholly yellow for half the length from the base and with three posterior yellow bands whose fore borders are undulating. Legs yellow. Wings pellucid; veins black, testaceous towards the base. Length of the body 5½ lines.

Dahleck.

127. LARRA ANNULATA. *Male.*—Black. Head silvery in front, yellow about the mouth. Antennæ ferruginous towards the tips. Thorax yellow along the fore border and with yellow tubercles; scutellum with two connected yellow spots on the fore border and with a yellow band on the hind border. Abdomen with five yellow bands, of which the third, fourth and fifth are successively more excavated in the middle of the hind border; under side with three yellow bands, of which the second and third are widely interrupted. Legs yellow; four anterior femora striped with black; hind femora black except at the tips. Wings pellucid; veins pale yellow, piceous towards the tips. Length of the body 3—4 lines.

Harkeko. Tajura.

Genus PALARUS, *Latr.*

128. PALARUS HUMERALIS, *Duf. Ann. Soc. Ent. Fr.* 1853, 379.

Cairo.

Inhabits Algeria.

Genus HELIORYCTES, *Smith.*

129. HELIORYCTES MELANOPYRUS, *Smith, Cat. Hym.* iv. 359, pl. 9, f. 3.

Harkeko.

Inhabits West Africa.

Genus CRABRO, *Fabr.*

130. CRABRO GRANULATUS. *Male.* — Black. Head very large, finely punctured; first and second points of the antennæ yellow. Thorax roughly punctured, with two yellow calli or tubercles on the fore border and one on each side; scutellum with a yellow point. Metathorax with a very slight keel. Abdomen with five yellow bands; first band excavated on the fore border; second slightly interrupted; third straightened in the middle; fourth and fifth

entire; under side with a yellow quadrate spot near the base. Legs yellow; femora towards the base and coxæ black; four posterior tarsi piceous towards the tips. Wings blackish; veins black. Length of the body 4½ lines.
Cairo.

131. CRABRO CONFINIS. *Male and female.*—Black. Head with silvery tomentum in front. First and second joints of the antennæ yellow. Thorax with two yellow calli on the fore border and with one on each side. Abdomen with three yellow bands; first and second widely interrupted; third entire; under side wholly black. Legs yellow; femora towards the base and coxæ black; tarsi with piceous tips. Wings blackish; veins black. Length of the body 2½—3 lines.
Cairo.

132. CRABRO PERPUSILLUS. *Male.* — Black, smooth, shining. Head with silvery tomentum towards the mouth. Tibiæ and tarsi piceous; hind tibiæ yellow towards the base. Wings blackish; veins black. Length of the body 2 lines.
Cairo.

Genus RHOPALUM, *Kirby.*

133. RHOPALUM FRATERNUM, *Smith.*
Cairo.

Genus CERCERIS, *Latr.*

134. CERCERIS TYRANNICA, *Smith, Cat. Hym.* iv. 447.
Dahlcck. Tajura.
Inhabits West Africa.

135. CERCERIS HISTRIONICA, *Klug, Symb. Phys.* pl. 47, f. 9.
Wády Gennéh. Wády Ferran. Mount Sinai.

136. CERCERIS EXCELLENS, *Klug, Symb. Phys.* pl. 47, f. 15.
Harkcko.

137. CERCERIS PULCHELLA, *Klug, Symb. Phys.* pl. 47, f. 14.
Harkcko. Tajura.

138. CERCERIS VIDUA, *Klug, Symb. Phys.* pl. 47, f. 11.
Harkeko. Tajura. Wády Ferran.

139. CERCERIS ALBOATRA. *Male.* — Black, roughly punctured. Front of the head with three large yellowish white spots; first and second spots elongated, longitudinal, with a yellowish white line

between them; third transverse, nearer the mouth. Antennæ piceous; first and second joints black; first yellowish white beneath. Thorax with three yellowish white calli, one on each side near the fore border; the third transverse on the hind border of the scutellum; tegulæ yellowish white. Abdomen with three yellowish white bands; first adjoining the hind border of the petiole; second in the middle: third subapical. Legs yellowish white; coxæ and femora black, the latter with yellowish white tips; hind tibiæ with a black streak on the outer side. Wings cinereous; tips blackish; veins black. Length of the body 4 lines. *Var. β.*—Petiole red. *Var. γ.*—Like *Var. β.* Femora red; hind femora with a black streak. Length of the body 4—5 lines.

The petiole is much more slender than that of C. vidua.

Wády Ferran.

140. CERCERIS CONTIGUA. *Male and female.*—Black, roughly punctured, with markings like those of C. alboatra. Antennæ red, piceous towards the tips. Petiole red. Legs yellowish white; femora red, striped with black, their tips yellowish white. Wings cinereous; tips blackish; veins black. Length of the body 4½ lines.

Tajura.

Genus PHILANTHUS, *Fabr.*

141. PHILANTHUS ABDELKADER, *St. Farg. Hym.* iii. 33.

Cairo. Heliopolis. Hor Tamanib.

142. PHILANTHUS DIADEMA, *Fabr. Ent. Syst.* ii. 289.

Cairo. Heliopolis.

Inhabits West Africa and South Africa.

143. PHILANTHUS MELLINIFORMIS, *Smith, Cat. Hym.* iv. 469.

Harkeko. Tajura. Wády Ferran. Mount Sinai.

Inhabits Sicily.

144. PHILANTHUS SULPHUREUS, *Smith, Cat. Hym.* iv. 469.

Wády Ferran.

Genus ZETHUS, *Fabr.*

145. ZETHUS FAVILLACEUS. *Male and female.* — Black, with hoary tomentum. Mouth testaceous. Antennæ and legs piceous. Petiole long, slender, cylindrical. Hind borders of the abdominal segments whitish. Wings cinereous; veins black. Length of the body 7 lines.

Tajura.

Genus EUMENES, *Latr.*

146. EUMENES CAFFRA. Vespa Caffra, *Linn. Syst. Nat.* i. 951.
Hor Tamanib. Massowah. Akeck. Harkeko. Tajura. Tòr.
Inhabits South Africa.

147. EUMENES TINCTOR. Sphex tinctor, *Christ. Hym.* 341, pl. 31,
f. 1.
Cairo. Massowah. Harkeko.
Inhabits West Africa.

148. EUMENES FENESTRALIS, *Sauss. Mon. Guêpes, Sol.* 53, pl. 10, f. 6.
Rafla.
Inhabits West Africa.

149. EUMENES DIMIDIATIPENNIS, *Sauss. Mon. Guêpes, Sol.* 51, 33.
Hor Tamanib. Massowah. Akeck. Harkeko. Rafla. Tajura.
Pharaoh's Baths. Wàdy Ferran. Tòr. Mount Sinai.
Inhabits Hindostan.

150. EUMENES ESURIENS. Vespa esuriens, *Fabr. Mant. Ins.* i. 39.
Massowah. Wàdy Ferran. Wàdy Nash.
Inhabits Hindostan.

151. EUMENES POMIFORMIS. Vespa pomiformis, *Rossi, Faun. Etr.*
85.
Cairo.
Inhabits South Europe.

152. EUMENES NIGRA, *Brullé, Hist. Nat. Il. Canar.* ii. 89.
Pharaoh's Baths. Mount Sinai.
Inhabits the Canary Isles.

153. EUMENES SAVIGNYI, *Guér. Icon. R. An.* 446, pl. 72, f. 4.
Cairo.
Inhabits West Africa.

154. EUMENES BISIGNATUS. *Male.*—Black, minutely punctured.
Head in front and mouth luteous. Antennæ reddish, with a broad
piceous band beyond the middle. Prothorax reddish; disk black,
including two reddish clavate stripes. Abdomen capitate, red; disk
black towards the tip; petiole very long and slender, black at the
base. Legs reddish. Wings dark cinereous; veins black. Fore

wings with a lurid costal streak and with an exterior blackish costal streak; veins black. Length of the body 6½ lines.
Wády Ferran.

155. EUMENES LEPTOGASTER. *Male.*—Black, with silvery tomentum, minutely punctured. Head in front yellow; mouth luteous. Antennæ reddish, with a blackish streak above beyond the middle. Thorax yellow; disk black; a reddish spot on each side adjoining the tegulæ, which are also reddish. Metathorax and abdomen reddish; petiole black at the base and with a black stripe; club black, with one or two yellow bands, of which the second is sometimes interrupted; under side with five yellow bands. Legs reddish. Wings cinereous; veins black. Fore wings with a blackish costal subapical patch. Length of the body 6½—7 lines.
Wády Ferran.

156. EUMENES SIGNICORNIS. Black, roughly punctured. Front yellow, slightly silvered: a small yellow dot between the antennæ; a short yellow line along the hind side of each eye and a shorter yellow line along the outer side of each eye. Antennæ reddish, with a piceous streak near the base and a piceous band near the tip. Thorax: a band on the fore border straightened in the middle tegulæ, two short bands on the scutellum, two transverse streaks on each side and a large spot on each side of the metathorax yellow. Legs reddish; tibiæ and four anterior femora yellow on the outer side. Wings cinereous; veins black. Fore wings with a blackish streak along the apical part of the costa. Length of the body 5? lines.
Massowah.

157. EUMENES DILECTULA. *Male.* — Black, finely punctured. Front yellowish white, with silvery tomentum. Antennæ tawny beneath; first joint yellowish white beneath. Thorax: an interrupted band on the fore border, tegulæ and two nearly connected dots on the scutellum yellowish white. Petiole red on each side of its hind part; a yellowish white band on its hind border. Abdomen with a yellowish white band near the tip. Legs yellowish white; middle femora on the inner side and hind femora except at the tips on the outer side black. Wings pellucid; veins black. Length of the body 3 lines.
Wády Gennéh.

Genus RHYNCHIUM, *Sauss.*

158. RHYNCHIUM OCULATUM. Vespa oculata, *Fabr. Ent. Syst.* ii. 265.
Cairo.
Inhabits South Europe.

159. RHYNCHIUM LATERALE. Vespa lateralis, *Fabr. Ent. Syst.* ii, 275.

Akeek.

160. RHYNCHIUM CYANOPTERUM, *Sauss. Mon. Guêpes, Sol. Suppl,* 177.

Hor Tamanib. Akeek.

Inhabits West Africa.

161. RHYNCHIUM ARDENS. *Female.*—Red, very finely punctured. Eyes piceous, deeply notched on the inner side. Third and following segments of the abdomen black. Wings blackish, cinereous and partly lurid towards the base, the latter hue most prevalent along the costa of the fore wings. Length of the body 7 lines. *Male?*—Head black above; a triangular yellowish white spot above the antennæ; front yellowish white, silky. Thorax with a large black spot on the scutum. Abdomen black at the base. Length of the body 6 lines.

Allied to B. cyanopterum; the outline of the luteous part of the fore wings is much more oblique.

Mount Sinai.

162. RHYNCHIUM FERVENS. *Male and female.*—Red, very finely punctured. Head black above; a triangular yellowish white spot above the antennæ, a streak of the same hue in the notch of each eye; front yellowish white, silky. Eyes piceous, deeply notched on the inner side. Thorax with a large black spot on the scutum. Abdomen black at the base; third and following segments black. Wings blackish, cinereous and partly lurid towards the base, the latter hue most prevalent along the costa of the fore wings. Length of the body 5—6 lines.

Mount Sinai.

163. RHYNCHIUM ZONATUM. *Male and female.* — Red, thickly punctured. Head yellow in front. First joint of the antennæ yellow beneath in the male. Second and third segments of the abdomen yellow. Wings cinereous, luteous along the costa; veins luteous, piceous towards the tips. Length of the body 5—7 lines.

Distinguished from B. brunneum and from R. Carnaticum by the markings of the abdomen and by the wings, whose tips are not brown.

Wàdy Ferran.

Genus ODYNERUS, *Latr.*

164. Odynerus bellatus, *Sauss. Mon. Guépes, Sol.* 210, pl. 18, f. 10.

Tajura.

Inhabits West Africa.

165. Odynerus parietum.   Vespa parietum, *Linn. Syst. Nat.* i. 949,

Cairo.

Inhabits Europe.

166. Odynerus crenatus, *St. Farg. Hym.* ii. 629.

Cairo.

167. Odynerus trimarginatus, *Zett. Ins. Lapp.* 456.

Wády Ferran.   Wády Nash.   Mount Sinai.

Inhabits Europe.

168. Odynerus parvulus, *St. Farg. Hym.* ii. 631.

Tajura,

Inhabits Algeria.

Synopsis of the following species of Odynerus :—

A. Abdomen with no red colour.

a. Abdomen yellow, except a black spot.   -   flavus.

b. Abdomen black, with bands.

   * Abdomen with white bands. -   -   -   - albifer.

  ** Abdomen with yellow bands.

    † Abdomen with the band of the first segment excavated in the disk.

    ‡ No spot behind the first band.

    § The first band slightly excavated.

    X Abdomen with two bands. -   -   -   fumipennis.

   XX Abdomen with three bands. -   -   -   disjunctus.

    §§ The first band deeply excavated.

    X Femora mostly red. -   -   -   -   inornatus.

   XX Femora yellow.

    o Wings cinereous. -   -   -   -   concinnulus.

    oo Wings blackish.

    ⟶ Band of the prothorax deeply excavated on the hind side. -   ·   -   -   -   rotundatus.

-+ -+ Band of the prothorax slightly excavated on the
 hind side. - - - - - guttulosus.
 ‡‡ A spot on each side behind the first band. -   cingulifer.
 †† Abdomen with the band of the first segment
  not excavated.
  ‡ First band with a yellow spot on each side in
   front. - · · - -   mutabilis.
  ‡‡ First band with no spot in front · ·   alienus.
 ✶✶✶ Abdomen with luteous bands.   ·
  † Band of the abdomen broadly excavated on the
   fore side. - - · - -   exustus.
  †† Band of the abdomen narrowly excavated on
   the fore side.   - . - -   stipatus.
 B. Abdomen with red colour.
 a. Abdomen with the band of the first segment
  excavated.
 ✶ Band very broadly excavated.
 † Scutellum with a band.
 ‡ Abdomen with a patch on each side behind the
  first band. - - - - -   dotatus.
 ‡‡ Abdomen with no patch. - - -   diversus.
 †† Scutellum with two spots. - - -   cognatus.
 ✶✶ Band broadly excavated.
 † Band of the prothorax not concisely abbreviated
  on each side. - - - -   tinctus.
 †† Band of the prothorax concisely abbreviated on
  each side. - - - - -   privatus.
 ✶✶✶ Band narrowly excavated. - - -   torridus.
 b. Abdomen with the first band not excavated. -   selectus.

169. ODYNERUS FLAVUS. *Male.* — Yellow, finely punctured.
Vertex black. Mandibles with black tips. Antennæ red; first
joint yellow. Disk of the fore part of the thorax, base of the
scutellum and a conical spot on the metathorax black. Abdomen
with a black basal spot. Wings slightly cinereous; veins black.
Fore wings with a narrow blackish border. Length of the body
5 lines.

Wâdy Ferran.

170. ODYNERUS TORRIDUS. *Female.*—Black, roughly punctured.
Head with a streak behind each eye, three spots between the eyes
and face red. First, second and base of the third joint of the antennæ
red. Fore part of the thorax, a band on the scutellum and a large
spot on each side of the scutellum red. Abdomen with two red
bands; first band very broad, very deeply excavated in the middle of

the fore border; second excavated on each side of the fore border.
Legs red. Wings cinereous; veins black. Fore wings with an
irregular blackish costal stripe. Length of the body 4½ lines.
Mount Sinai.

171. ODYNERUS TINCTUS. *Female.* — Black, finely punctured.
Head with a red streak behind each eye, a red spot in the notch of each
eye and face red. First and second joints of the antennæ red.
Fore part of the thorax and a large spot on each side of the metathorax red. First segment of the abdomen red, yellow along the
hind border and including a very large black spot; hind borders of
the other segments yellow. Legs red. Wings slightly cinereous;
veins black, tawny at the base. Fore wings with an irregular and
diffuse blackish streak along the costa. Length of the body 5 lines.
Wâdy Gennèh.

172. ODYNERUS SELECTUS. *Female.* — Black, finely punctured.
Face and first joint of the antennæ red. A band on the fore part of
the thorax with its hind border excavated in the middle, a spot near
each tegula, two almost connected spots on the scutellum and a spot
on each side of the metathorax yellow; hind border of the scutellum
red. First segment of the abdomen red, bordered with yellow; hind
border of the second segment yellow. Legs red; coxæ black. Wings
slightly cinereous; veins black, reddish at the base. Length of the
body 3¼ lines.
Mount Sinai.

173. ODYNERUS PRIVATUS. *Male.*—Black, punctured. Head with
a triangular spot above the antennæ, a streak in the notch of each
eye and face yellowish white. First and second joints of the antennæ
luteous. Thorax with an interrupted luteous band on the fore border
and with a luteous spot on each side of the metathorax. Abdomen
with a very broad luteous band near the base, this band very deeply
excavated in the middle of its fore side; second segment with a
luteous dot on each side in the disk, its hind border, like that of the
third segment, yellow. Legs yellow; coxæ and hind femora black.
Wings cinereous; veins black. Fore wings with a diffuse blackish
stripe along the costa. Length of the body 3½—4 lines.
Wâdy Gennèh. Wâdy Ferran.

174. ODYNERUS DIVERSUS. *Male.*—Black, thickly punctured.
Head with a triangular spot above the antennæ, a notch in eye and
the face yellowish white. First and second joints of the antennæ
luteous. Fore border of the prothorax with a broad luteous band,
which is deeply excavated in the middle of the hind side; scutellum

luteous along the hind border; a large luteous spot on each side of the metathorax. Abdomen with a luteous band near the base, this band slightly excavated on its fore side; hind borders of the second and third segments pale yellow. Legs yellow; hind femora piceous above. Wings slightly cinereous; veins black, tawny at the base. Length of the body 4 lines.

Wâdy Gennêh.

175. Odynerus dotatus. *Male.* — Black, thickly punctured. Head with three yellowish white angular spots, one above the antennæ and one in the notch of each eye; face yellowish white. First and second joints of the antennæ luteous; first yellow beneath. Fore part of the thorax with a broad yellow partly luteous band, which is very deeply excavated on the hind side; scutellum with a luteous band and a luteous hind border; metathorax with a large luteous patch on each side. First segment of the abdomen luteous, with a black disk in front and with a yellow hind border; two posterior yellow bands connected on each side; first band very widely interrupted in the middle; second undulating along its fore side; under side yellow, with a black dot on each side and with three black bands. Legs pale luteous. Wings slightly cinereous; veins black, testaceous at the base. Fore wings with an irregular blackish streak along the costa. Length of the body 4½ lines.

Wâdy Gennêh.

176. Odynerus cognatus. *Male.* — Black, thickly punctured. Head with a luteous line along the hind border of each eye; a triangular spot above the antennæ, a streak in the notch of each eye and face yellowish white. First and second joints of the antennæ luteous, Prothorax with a luteous band on the fore border, a spot on each side near the tegula, an interrupted band on the scutellum and a patch on each side of the metathorax luteous. Abdomen with the first segment luteous, excepting its black. fore disk; hind borders of the following segments yellow. Legs yellow; femora above and coxæ black. Wings cinereous; veins black. Fore wings with a blackish streak along the costa. Length of the body 4 lines.

Wâdy Ferran.

177. Odynerus exustus. *Female.* —Black, thickly punctured, with red markings; a transverse streak behind each eye, an angular spot behind the antennæ, one in the notch of each eye, face, a band on the fore part of the thorax with a deeply excavated hind border, a dot near each tegula, a band on the scutellum, a large spot on each side of the metathorax and two abdominal bands with excavated fore borders. Legs and first joint of the antennæ red. Wings dark cinereous; veins black, red at the base. Fore wings blackish along the costa. *Male.*—Markings of the head yellow, those of the thorax and of the abdomen luteous. Abdomen with three bands and with a posterior spot in the disk. First joint of the antennæ yellow beneath. Legs luteous. Length of the body 4½—5 lines.

Mount Sinai.

178. ODYNERUS STIPATUS. *Male.* — Black, thickly punctured; markings luteous. A transverse streak behind each eye, face, a band on the fore part of the thorax with a deeply excavated hind border, a spot near each tegula, a band on the scutellum, a large spot on each side of the metathorax, and two abdominal bands; the bands deeply excavated on the fore border, the hinder one with the excavation extending obliquely on each side. Legs and first joint of the antennæ luteous. Wings blackish; veins black. Length of the body 3 lines.

Cairo?

179. ODYNERUS ALBIFER. *Female.* — Black, thickly punctured. Head with a narrow white dot behind each eye and with two connected white dots between the antennæ. Antennæ with a white point on the first joint. Thorax with an angular white spot on each side in front; a triangular white dot on the proscutellum. Abdomen with five white bands; fore side of the first band undulating, notched in the middle; second band widened on each side; fourth undulating along the fore border; fifth more undulating. Tibiæ white on the upper side; tarsi reddish towards the tips. Wings blackish cinereous; veins black. Fore wings blackish along the costa. Length of the body 5 lines.

Cairo.

180. ODYNERUS GUTTULOSUS. *Female.*—Black, thickly punctured; markings pale yellow. A transverse streak behind each eye, another in each notch of the eye, a transverse spot above the antennæ, face, a band on the fore part of the thorax, a large spot near each tegula, two almost contiguous spots on the scutellum, two posterior transverse almost connected streaks, a large spot on each side of the metathorax, and six abdominal bands; first abdominal band much dilated on each side; second connected on each side with a large forked mark; third, fourth, fifth and sixth connected on each side with an oblique streak. First joint of the antennæ yellow beneath and towards the base.

Legs yellow. Wings blackish; veins black, tawny at the base.
Length of the body 4—4½ lines.
Harkeko.

181. ODYNERUS CINGULIFER. *Female.*—Black, thickly punctured;
markings yellow. A streak behind each eye, another in the notch of
each eye, a transverse spot above the antennæ, face, a band on the
fore part of the thorax much dilated on each side, a large spot near
each tegula, two spots on the scutellum, a posterior abbreviated band,
a large spot on each side of the metathorax, six abdominal bands, the
first and second each connected with a patch on each side, two ventral
bands, the first broad, much excavated in the middle of its hind side.
First joint of the antennæ yellow beneath. Legs yellow; femora striped
above with black. Wings blackish, cinereous; veins black. Fore wings
with a blackish streak along the apical part of the costa. *Var. β.*—Spot
above the antennæ very small, not connected with the face ; first and
second abdominal bands not connected with the patches on each side ;
first ventral band merely represented by a dot on each side. Length
of the body 4½—5 lines.
Harkeko.

182. ODYNERUS INORNATUS. *Female.*—Black, thickly and minutely
punctured ; markings yellow. A streak behind each eye, another in
the notch of each eye, a spot above the antennæ, face including an
angular black spot, a very narrow band on the fore part of the thorax
much dilated on each side, a spot near each tegula, a lunate abbreviated band, a large spot on each side of the metathorax, six
abdominal bands, two ventral bands; first abdominal band broad,
much excavated in the middle of its fore border; second slightly
notched on each side of its fore border; third and following much
abbreviated ; second ventral band with three excavations in its fore
border. Mandibles red. Antennæ red towards the base; first joint
with a black streak, yellow beneath. Legs dark red; femora and
tibiæ striped with yellow. Wings blackish cinereous ; veins black.
Fore wings blackish along the costa. Length of the body 5½ lines.
Cairo.

183. ODYNERUS ROTUNDATUS. *Male.*—Black, short, thickly and
minutely punctured; markings yellow. A short streak behind each
eye, a streak in the notch of each eye, a conical spot above the
antennæ in connection with the face, a band on the fore part of the
thorax excavated in the middle of the hind border, a spot near each
tegula, two posterior much abbreviated bands, an elongated spot on

each side of the metathorax, six abdominal bands, one ventral band; first abdominal band broad, excavated on its fore border; second broad; third, fourth and fifth narrow; sixth abbreviated, ventral band postmedial, undulating along its fore side. Antennæ tawny beneath; first joint yellow, with a black streak near its tip above. Legs yellow; coxæ black. Wings blackish; veins black. Length of the body 2¾ lines.

Cairo.

184. ODYNERUS FUMIPENNIS. *Male and female.*—Black, thickly and minutely punctured; markings yellow. A broad streak behind each eye, another in the notch of each eye, a triangular spot above the antennæ, face, a band on the fore part of the thorax much excavated along its hind border, a spot near each tegula, a band on the scutellum, two broad abdominal bands; one ventral band; first abdominal band excavated in the middle of its fore border; second slightly notched on each side of its fore border; ventral band notched on each side of its fore border. First joint of the antennæ yellow beneath and at the base. Legs yellow; coxæ black; femora striped with black above. Wings blackish; veins black. *Var. β.*—Antennæ red beneath. Streaks of the head narrow; spot connected with the face; band on the scutellum and a posterior band abbreviated; second abdominal band not notched; femora wholly yellow except at the base above. Length of the body 3½—4 lines.

Cairo.

185. ODYNERUS CONCINNULUS. *Male.*—Black, thickly and minutely punctured; markings yellowish white. A slender streak behind each eye, another in the notch of each eye, a triangular spot in connection with the face, a band on the fore part of the thorax much excavated on its hind side, a spot near each tegula, a band on the scutellum and a posterior band both abbreviated, six abdominal bands, one ventral band; first abdominal band much excavated in the middle of its fore border; second undulating along its fore border; third merely represented by a dot on each side; fourth, fifth and sixth abbreviated; ventral band undulating along its fore border. Antennæ tawny beneath; first joint yellowish white. Legs yellowish white. Wings blackish; veins black. Length of the body 3 lines.

Dahleck.

186. ODYNERUS NOTABILIS. *Male.*—Black, thickly and minutely punctured; markings pale yellow. A slender streak behind each eye, another in the notch of each eye, a triangular spot in connection with the face, a band on the fore part of the thorax much excavated in the

middle of the hind border, a large spot near each tegula, two spots on
the scutellum, a narrow abbreviated posterior band, a large spot on
each side of the metathorax, six abdominal bands, six ventral bands,
a spot before and a patch behind each side of the first abdominal
band; first ventral band notched in the middle of its hind border;
third, fourth, fifth and sixth ventral bands very widely interrupted.
First joint of the antennæ pale yellow beneath.   Legs pale yellow;
femora above at the base and trochanters black.   Wings blackish;
veins black, tawny at the base.   Length of the body 4½ lines.

Dahleck.

187. ODYNERUS DISJUNCTUS.   *Female.* — Black, thickly and
minutely punctured; markings yellow.   A broad streak behind each
eye, another in the notch of each eye, a triangular spot in connection
with the face, a broad band on the fore part of the thorax deeply
excavated on its hind side, a spot near each tegula, an abbreviated
band on the scutellum, a narrower posterior abbreviated band, a large
spot on each side of the metathorax, three abdominal bands, one
ventral band : first abdominal band broad, excavated in the middle of
the fore border; second and ventral band broader on each side.   First
joint of the antennæ yellow, with a black streak above at the tip.
Legs yellow; femora at the base, trochanters and coxæ black.   Wings
blackish; veins black.   Length of the body 4 lines.

Cairo.

188. ODYNERUS ALIENUS.   *Male.*—Black, thickly and minutely
punctured; markings yellowish white.   A slender streak in the notch
of each eye, face, narrow bands on the hind borders of the abdominal
and ventral segments from the first to the fourth.   First joint of the
antennæ yellow beneath.   Legs tawny; hind femora above and coxæ
black.   Wings cinereous; veins black, tawny at the base.   Fore
wings blackish along the costa.   Length of the body 3 lines.

Allied to O. minutus; thorax wholly black in front.

Wâdy Gennèh.

Fam. VESPIDÆ, *Steph.*
Genus BELENOGASTER, *Sauss.*

189. BELENOGASTER JUNCEUS.   Vespa juncea, *Fabr. Ent. Syst.* ii.
277.

Cairo.

Inhabits West Africa.

Genus ICARIA, *Sauss.*

190. ICARIA CINCTA. Epipona cincta, *St. Farg. Hym.* i. 541.
Dahleck. Tajura.
Inhabits West Africa.

Genus POLISTES, *Latr.*

191. POLISTES GALLICA. Vespa Gallica, *Linn. Syst. Nat.* i. 949.]
Cairo. Heliopolis.
Inhabits South Europe.

192. POLISTES MARGINALIS. Vespa marginalis, *Fabr. Ent. Syst.*
ii. 264.
Harkeko.
Inhabits West Africa and South Africa.

Genus VESPA, *Linn.*

193. VESPA ORIENTALIS, *Linn. Syst. Nat. Mant.* 540.
Cairo. Shoobra. Dahleck. Wâdy Gennèh. Pharoah's Baths.
Wâdy Ferran. Wâdy Nash. Wâdy Hebran. Mount Sinai.
Inhabits South Europe and Hindostan.

Tribe ANTHOPHILA.
Fam. ANDRENIDÆ.
Genus COLLETES, *Latr.*

194. COLLETES SUCCINCTA. Apis succincta, *Linn. Syst. Nat.* i.
955.
Cairo.
Inhabits Europe.

Genus PROSOPIS, *Fabr.*

195. PROSOPIS ALBONOTATA. *Male and female.*—Black, shining.
Front wholly or on each side, fore border of the thorax, a dot on each
side of the thorax, tegulæ, tibiæ and tarsi pure white. Antennæ
tawny beneath; first joint white beneath. Wings slightly cinereous;
veins black. Length of the body 2—3 lines.

It is distinguished from P. annulata by the white tibiæ.
Cairo. Tajura.

196. PROSOPIS NIGRITULA. *Female.* — Black, shining, with cinereous hairs. Wings blackish cinereous; veins black. Length of the body 2¼ lines.

The wholly black tibiæ distinguish this species from P. annulata and from P. affinis. It is more slender than P. anthracina.

Mount Sinai.

197. PROSOPIS RUFOCINCTA. *Male.*—Black, shining. Fore border of the thorax, femora at the tips and tibiæ and tarsi at the base whitish. Abdomen with a red band. Wings cinereous; veins black. Length of the body 3 lines.

It may be a variety of P. rubicola.

Wâdy Ferran.

Genus SPHECODES, *Latr.*

198. SPHECODES AFRICANUS, *St. Farg. Hym.*

Cairo.

Genus HALICTUS, *Latr.*

199. HALICTUS QUADRISTRIGATUS, *Latr. Hist. Nat. Ins.* xiii. 365.

Cairo.

Inhabits Europe.

200. HALICTUS PARVULUS. Apis parvula, *Fabr. Ent. Syst. Supp.* 277.

Wâdy Ferran.

Inhabits South Europe.

201. HALICTUS JUCUNDUS, *Smith, Cat. Hym.* i. 56.

Cairo.

Inhabits West Africa and South Africa.

202. HALICTUS DETERMINATUS. *Female.* — Black, with whitish hairs, thickly and very minutely punctured. Abdomen elliptical, with a reddish band on the hind border of each segment. Tarsi towards the tips and calcaria tawny. Wings cinereous, darker along the exterior border; veins black; stigma piceous. Length of the body 4 lines.

Mount Sinai.

G

203. **HALICTUS NIGRINUS.** *Female.* — Black, thickly and very minutely punctured, with cinereous hairs. Metathorax rather roughly punctured. Abdomen with a band of whitish hairs on the hind border of each segment; hind borders of the ventral segments piceous. Wings cinereous; veins and stigma black. Length of the body 4 lines.

Cairo.

204. **HALICTUS DISTINCTUS.** *Female.* — Black, thickly and very minutely punctured, with hoary hairs. Abdomen with a hoary band on the hind border of each segment. Tarsi reddish towards the tips. Wings cinereous, darker along the exterior border; veins black; stigma piceous. Length of the body 4½ lines.

Wâdy Gennéb.

205. **HALICTUS TIBIALIS.** *Male.* — Black, thickly and very minutely punctured, with cinereous hairs. Antennæ tawny beneath. Clypeus prominent, with a dingy whitish band on the fore border. Abdomen with a reddish piceous band and with cinereous hairs on the hind border of each segment. Four anterior femora sometimes striped with tawny; tibiæ and tarsi pale yellow, the former broadly striped with black on each side. Wings cinereous, darker along the exterior border; veins black, testaceous at the base; stigma piceous. Length of the body 4—4½ lines.

Longer than H. longulus; has much more resemblance to H. productus, a West African species, but the different colour of the legs distinguishes it from the latter.

Wâdy Ferran. Mount Sinai.

206. **HALICTUS DECORUS.** *Female.* — Black, shining, thickly and extremely minutely punctured, with whitish hairs. Antennæ red beneath towards the tips. Disk of the metathorax flat, with a slight longitudinal furrow. Abdomen elliptical, with well-defined white bands on the hind borders of the segments; hind borders of the ventral segments reddish. Tarsi tawny. Wings cinereous; veins black; stigma piceous. Length of the body 4 lines.

Allied to H. albescens of Hindostan, but the latter has four bands on the abdomen.

Harkeko.

Genus NOMIA, *Latr.*

207. NOMIA CRASSIPES. Eucera crassipes, *Fabr. Ent. Syst. Supp.* 278.
Rafla.
Inhabits Hindostan.

208. NOMIA OXYBELOIDES, *Westw. MSS*
Tajura.
Inhabits Hindostan.

209. NOMIA TEGULATA, *Westw. MSS.*
Inhabits Natal and Sierra Leone. A variable species.

210. NOMIA ZONARIA. *Male.*—Black, stout. Head and sides of the thorax with silvery whitish hairs. Antennæ red beneath and at the tips; first joint red. Thorax with cinereous pubescence. Abdomen with five white bands, which have a yellowish tinge in the middle. Legs tawny, thick; femora and tibiæ with white tomentum on the outside. Wings cinereous, broadly blackish along the exterior border; veins black, stigma piceous. Length of the body 6 lines.

The thin forked edge which forms the hind border of the scutellum distinguishes this and the following species from the other Nomiæ.
Harkeko.

211. NOMIA VESPOIDES. *Male.*—Black. Head and thorax with cinereous pubescence. Antennæ and disk of the scutellum red. Abdomen dark red, with three black bands and with three pale luteous bands, which are white at each end. Legs tawny, thick. Wings lurid, broadly blackish along the exterior border; veins black; stigma piceous. Length of the body 5½ lines.
Massowah.

212. NOMIA BICOLORIPES. *Female.*—Black. Head and thorax with cinereous pubescence. Antennæ dark red beneath. Hind borders of the abdominal segments lurid, with whitish tomentum. Legs tawny; four anterior femora and tibiæ black. Wings cinereous, broadly blackish along the exterior border; veins and stigma black. Length of the body 4 lines.

The bands of the abdomen are rather narrower than those of N. fulvohirta, a native of Sierra Leone.
Massowah.

213. NOMIA RUFIVENTRIS. *Female.*—Black, with whitish pubescence. Antennæ dark red. Abdomen red. Legs thickly clothed with yellowish hairs; tibiæ and tarsi red. Wings cinereous; veins black, testaceous at the base; stigma lurid. Length of the body 2½ lines.

Tajura.

214. NOMIA FEMORALIS. *Male.*—Black, with hoary pubescence. Front with shining white pubescence, including a transverse yellow spot in the disk. Antennæ tawny beneath, except at the base. Abdomen with two narrow bands and the tip chalybeous; under side thickly clothed with shining hoary hairs. Tarsi testaceous; hind femora much incrassated, red towards the tips; four anterior tibiæ striped with red; hind tibiæ red. Wings cinereous; veins and stigma black, the former tawny at the base. Length of the body 3½ lines.

The clothing and markings distinguish it from N. crudelis, *Westw.*, a Cape species.

Dahleck.

215. NOMIA AMPLA. *Female.*—Black, broad, with white pubescence. Antennæ red beneath, excepting the first and second joints. Tegulæ white, very large, with a blackish patch on the inner side. Abdomen with white bands. Legs with cinereous hairs. Wings cinereous; veins black, testaceous at the base; stigma piceous. Length of the body 5 lines.

Most allied to N. difformis, an inhabitant of Dalmatia and of the Crimea.

Tajura.

216. NOMIA SCRIPTIFRONS. *Female.*—Black, thickly and minutely punctured, with cinereous pubescence. Front with a broadly U-shaped white mark. Antennæ red beneath, except the first and second joints. Scutellum with a white spot on each side and a white hind border. Abdomen with a chalybeous band, which is clothed with cinereous pubescence on the hind border of each segment; under side piceous. Legs hairy; tibiæ and tarsi tawny. Wings pellucid; veins tawny; stigma testaceous. Length of the body 4 lines.

It may be the female of N. cinerascens, *Westw.*, a Natal species.

Massowah.

217. NOMIA EBURNEIFRONS. *Female.* — Black, thickly and minutely punctured, with hoary pubescence. Head yellowish white

and with white pubescence between the base of the antennæ and the mouth. Eyes reddish, broad. Antennæ red, subclavate; first joint white. Abdomen with a chalybeous tawny band on the hind border of each segment. Legs testaceous ; furrows black, except at the tips ; tibiæ with a piceous streak. Wings pellucid ; veins and stigma testaceous ; costal vein black, except at the base. Length of the body 4¼ lines.

Tajura.

218. Nomia pallicornis. *Male.*—Black, broad, thickly and minutely punctured, with whitish pubescence. Mouth testaceous. Antennæ tawny, luteous beneath. Metathorax largely punctured ; disk flat, with a slight longitudinal furrow. Abdomen with a testaceous streak on each side at the base. Tibiæ tawny, with a broad blackish stripe ; tarsi yellow, piceous towards the tips ; hind legs very thick. Wings pellucid ; veins and stigma testaceous, the former black towards the tips. Length of the body 4 lines.

Tôr.

Genus ANDRENA, *Fabr.*

219. Andrena cirtana, *Lucas, Expl. Sci. Alg.* iii. 178, pl. 6, f. 7. Wády Ferran. Mount Sinai. Inhabits Algeria.

220. Andrena dorsata. Melitta, *Kirby, Mon. Ap. Angl.* ii. 144. Cairo. Inhabits Europe.

221. Andrena fulvicrus. Melitta fulvicrus, *Kirby, Mon. Ap. Angl.* ii. 138. Cairo. Inhabits Europe and North Hindostan.

222. Andrena partita. *Female.*—Black, thickly and minutely punctured, with hoary pubescence. Antennæ red beneath, except towards the base. Abdomen red, with four blackish bands and with a fringe of white pubescence on the hind border of each segment ; first and second bands narrow ; third and fourth broad. Tarsi and hind tibiæ tawny ; hind legs fringed with long yellowish hairs. Wings pellucid ; veins testaceous, black towards the costa ; stigma piceous. Length of the body 4½ lines.

223. ANDRENA TURBIDA. *Female.*—Black, thickly clothed with hoary hairs. Abdomen with hoary hairs on the hind borders of the segments; under side without hairs. Tibiæ and tarsi with testaceous hairs. Wings cinereous; veins black; stigma piceous. Length of the body 4 lines.

Mount Sinai.

224. ANDRENA DISPARILIS. *Female.* — Æneous, thickly and minutely punctured, with cinereous pubescence. Eyes and antennæ black. Abdomen red, a little broader than the thorax. Tibiæ, tarsi and tips of the femora testaceous. Wings pellucid; veins and stigma pale testaceous. Length of the body 2¾ lines.

Cairo.

225. ANDRENA MUNDA. *Male.*—Black. Head, thorax, legs and base of the abdomen clothed with white hairs. Abdomen with four bands of white pubescence on the hind borders of the segments, these bands narrower on the under side; hairs brown above at the tip. Wings cinereous; veins black; stigma piceous. Length of the body 5 lines.

Cairo.

226. ANDRENA BREVIPENNIS. *Female.*—Black. Head and thorax with whitish hairs. Face yellow, smooth, shining, not pilose. Antennæ red, short; first joint yellow. Abdomen red, black towards the tip; a band of yellowish white tomentum on the hind border of each segment; a black spot above on each side of the second segment; two or three yellow patches on each side beneath. Legs yellow. Wings pellucid, short: veins yellow. Length of the body 4½—5 lines.

Harkeko. Tajura.

227. ANDRENA VENUSTA. *Male.*—Black. Head and legs thickly clothed with pale luteous hairs. Head with white hairs on each side beneath. Abdomen with a band of luteous pubescence on the hind border of each segment. Wings cinereous; veins black; stigma piceous. Length of the body 5—5½ lines.

Cairo.

Genus OSMIA, *Panz.*

228. OSMIA EMARGINATA, *St. Fary. Hym.* ii. 317.

Mount Sinai.

Inhabits France.

229. Osmia melanogaster, *Spin. Ins. Lig.* ii. 63.
Cairo.
Inhabits South Europe.

230. Osmia ænea. Apis ænea, *Linn. Syst. Nat.* i. 955.
Cairo.
Inhabits Europe and the Canary Isles.

231. Osmia contracta. *Male and female.*—Black, stout. Head and thorax with white hairs. Mandibles red; tips black. Abdomen with five bands of white tomentum; a white spot on each side between the first band and the base. Legs with white tomentum. Wings cinereous; veins and stigma black. Length of the body 3½ lines.
Harkeko. Dahleck.

Genus MEGACHILE, *Latr.*

232. Megachile argentata. Apis argentata, *Fabr. Ent. Syst.* ii. 336.
Massowah. Harkeko. Wády Ferran. Mount Sinai.
Inhabits Europe and Algeria.

233. Megachile xanthopus, *Gerst. Verh. Ak. Berl.* 1855.
Harkeko.

234. Megachile gratiosa, *Gerst. Verh. Ak. Berl.* 1855.
Harkeko.

235. Megachile fulvescens. *Male and female.*—Black. Head and thorax thickly covered with cinereous hairs. Abdomen tawny at the base; hind borders of the segments with bands of testaceous pubescence, except towards the tip, where the bands are more or less cinereous; under side tawny, densely clothed in the female with yellow hairs; tip of the male with six short spines, the two outer pairs very minute. Legs with hoary hairs in the male, with tawny hairs in the female. Wings cinereous; veins black. Length of the body 3½—4½ lines.
Harkeko. Wády Gennèh. Wády Ferran. Mount Sinai.

236. Megachile adusta. *Female.*—Black, thickly clothed with short ochraceous hairs. Head with a triangular spot of luteous tomentum on the face; hind part with white hairs on each side.

Mandibles and antennæ red, with black tips. Thorax with a tuft of white hairs by each tegula; metathorax with white hairs on each side. Legs red. Wings luteous, with tawny veins; exterior half blackish, with black veins. Length of the body 5½ lines.

Has most resemblance to M. rufipes, of West Africa, but differs much from that species in the abdomen and in the wings.

Akeek.

237. MEGACHILE INORNATA. *Male and female.*—Black. Head, thorax and legs clothed with cinereous hairs. Hind borders of the abdominal segments with bands of cinereous pubescence in the male and of tawny pubescence in the female ; under side of the abdomen of the female thickly clothed with whitish hairs ; tip of the male serrated, deeply notched in the middle. Tarsi with tawny tips ; fore femora of the male tawny beneath and at the tips ; fore tibiæ and fore tarsi of the male pale yellow. Wings cinereous, blackish along the exterior border ; veins black, tawny at the base ; stigma piceous. Length of the body 5—6 lines.

Allied to M. sericans, but the colour of the abdomen beneath is different.

Mount Sinai.

238. MEGACHILE INFICITA. *Male.* — Head and thorax clothed with whitish hairs. Head with yellowish hairs on the front. Mandibles red ; tips black. Abdomen with four bands of white pubescence ; tip white, with six short spines. Legs red ; femora and four posterior tibiæ black above. Wings cinereous ; veins and stigma black. Length of the body 4 lines.

Wâdy Ferran.

239. MEGACHILE CONFICITA. *Male.*—Black, thickly clothed with yellowish cinereous hairs. Antennæ dark red beneath. Abdomen with a whitish band on the hind border of each segment; tip with six short spines. Legs red ; femora and tibiæ striped with black. Wings cinereous, broadly blackish along the exterior border ; veins and stigma black. Length of the body 4 lines.

Cairo.

240. MEGACHILE DESPECTA. *Male and female.*—Black. Head and thorax clothed with whitish hairs. Abdomen with a band of testaceous tomentum on the hind border of each segment; under side reddish, with bands of whitish pubescence; tip with six small spines. Legs red ; fore femora black beneath. Wings cinereous, broadly

bordered with blackish; veins black, tawny towards the base; veins tawny. Length of the body 3½—5 lines.
Wâdy Gennêh. Wâdy Ferran.

Genus CHALCEODOMA, *St. Farg.*

241. CHALCEODOMA SICULA. Apis sicula, *Rossi, Mant. Fn. Etr. App*, ii. 139, pl. 4, f. D.
Cairo. Heliopolis.
Inhabits South Europe and the Canary Isles.
"Constructs long mud nests."—*Lord.*

Genus DIOXYS, *St. Farg.*

242. DIOXYS CHALCEODA, *Lucas, Expl. Sci. Alg.* iii. 207, pl. 7, f. 6.
Heliopolis.
Inhabits Algeria.

Genus STELIS, *Panz.*

243. STELIS MINUTUS, *St. Farg. Enc. Meth.* x. 481.
Cairo.
Inhabits Europe.

244. STELIS DIMIDIATUS. *Male and female.*—Black. Head and thorax thickly clothed with short cinereous hairs. Mouth much more than half the length of the body. Abdomen red. Legs with cinereous hairs. Wings cinereous; veins black; stigma piceous. *Female.*—Abdomen above with black spots on each side and with a subapical black band, beneath with interrupted black bands. Wings cinereous; veins black; stigma piceous. Length of the body 3½ lines.
Hor Tamanib.

Genus ANTHIDIUM, *Fabr.*

245. ANTHIDIUM TESSELLATUM, *Klug, Symb. Phys.* iii. pl. 28, f. 4.
Wâdy Hebran. Mount Sinai.

246. ANTHIDIUM PULCHELLUM, *Klug, Symb. Phys.* iii. pl. 28, f. 11.
Wâdy Ferran. Wâdy Hebran. Mount Sinai.

247. ANTHIDIUM SUBOCHRACEUM. *Male.*—Tawny. Head with white hairs, yellow in front; vertex black. Mandibles with black tips.

H

Antennæ reddish. Thorax and pectus black, with tawny hairs; a line along each side of the thorax and hind border of the scutellum tawny. Abdomen thickly and minutely punctured. Wings cinereous, broadly blackish along the exterior border; veins black, tawny at the base; stigma piceous. *Var. β.*—Abdomen with three black bands. Length of the body 4—4½ lines.

Mount Sinai.

248. ANTHIDIUM SIGNIFERUM. *Male.*—Black, thick. Head and thorax with short slightly gilded hairs. Head with a transverse interrupted yellow line on the hind border; fore part testaceous, with whitish hairs. Antennæ dark red beneath; first joint yellow. Thorax with a yellow stripe on each side and with four yellow spots on the hind border of the scutellum. Abdomen with six yellow bands on the hind borders of the segments; bands from the first to the fourth notched on each side of the fore border and dilated at each end; fifth with a transversely elongated black dot on each side. Legs yellow; femora striped with red; tarsi red. Wings blackish cinereous, broadly blackish along the exterior border; veins and stigma black. Length of the body 5 lines.

Most allied to A. Latreillii, a European species.

Hor Tamanib.

Genus CERATINA, *Latr.*

249. CERATINA MAURITANICA, *St. Farg. Hym.* ii. 507.

Cairo.   Tajura.

Genus ALLODAPE, *St. Farg.*

250. ALLODAPE SYRPHOIDES. *Female.*—Æneous-green. Clypeus yellow. Eyes red. Antennæ black, tawny beneath; first joint yellow beneath. Abdomen yellow, with four black dorsal bands. Legs yellow; hind femora black, with yellow tips; hind tibiæ with a black streak on each side. Wings pellucid; veins and stigma pale yellow. Length of the body 2½ lines.

Tajura.

Var. Female.—Tawny. Head yellow in front. Eyes piceous. Antennæ piceous; first joint tawny. Thorax with three piceous stripes; scutellum yellow. Metathorax black. Abdomen yellow, with five black bands. Legs pale tawny. Wings pellucid; veins and stigma pale yellow. Length of the body 2½ lines.

Tajura.

Genus NOMADA, *Fabr.*

251. Nomada variabilis, *Lucas, Expl. Sci. Alg.* iii. 216, pl. 10, f. 7.
Cairo.

252. Nomada pusilla, *St. Farg. Hym.* iii. 466.
Cairo.

253. Nomada agrestis, *Fabr. Ent. Syst.* ii. 347.
Cairo.
Inhabits South Europe.

Genus EPEOLUS, *Latr.*

254. Epeolus nigriventris. *Female.* — Black, with white tomentum. Eyes reddish. Antennæ black; first joint tawny. Disk of the thorax piceous. Abdomen yellow, with four black dorsal bands; third and fourth bands very narrow; under side yellow. Wings cinereous, blackish along the exterior border; veins and stigma black. Length of the body 3 lines.
Tajura.

Genus CŒLIOXYS, *Latr.*

255. Cœlioxys caudata, *Spin. Ann. Soc. Ent. Fran.* vii. 535.
Mount Sinai.
Inhabits South Europe and Egypt.

256. Cœlioxys antica. *Male.*—Black. Head, except the vertex, pectus and legs thickly covered with silvery white tomentum. Front slightly tinged with yellow. Eyes lurid. Antennæ tawny beneath. Thorax with luteous speckles, with four short luteous stripes and with a posterior abbreviated luteous band. Abdomen with eight yellowish white bands; first, fifth and seventh bands very narrow; under side with three silvery white bands, which are narrowest in the middle; tip dark red, with four smaller spines above and with one on each side. Wings cinereous, darker along the exterior border; veins blaek; stigma piceous. Length of the body 4 lines.
Harkeko.

257. Cœlioxys rufispina. *Male.*—Black. Head with white hairs. Eyes lurid. Mandibles red; tips black. Mouth red.

Antennæ dark red beneath. Thorax with two white spots in front
and four behind. Abdomen with six narrow white bands and with a
white subapical spot; under side with four broader white bands;
tips red, with two rather long spines and with six smaller spines.
Legs with white tomentum; fore tibiæ and fore tarsi dark red.
Wings cinereous, broadly blackish along the exterior border; veins
and stigma black. Length of the body 3½ lines.
Harkeko.

Genus MELECTA, *Latr.*

258. MELECTA PLURINOTATA, *Brullé, Expl. Sci. Morée, Zool.* 343.
Cairo.
Inhabits South Europe and West Asia.

Genus CROCISA, *Jurine.*

259. CROCISA SCUTELLARIS. Nomada scutellaris, *Fabr. Ent. Syst.*
ii. 346.
Cairo. Harkeko. Mount Sinai.
Inhabits Europe, Siberia, Ceylon.

Genus EUCERA, *Scopoli.*

260. EUCERA FULVESCENS. *Male.*—Black, thickly clothed with
ochraceous hairs. Clypeus yellow. Eyes livid. Antennæ somewhat shorter than the body. Abdomen with black hairs towards the
tip. Legs with tawny and cinereous hairs; tarsi red beneath.
Wings cinereous; veins black; stigma piceous. Length of the body
6 lines.
Cairo.

261. EUCERA CINERASCENS. *Male.*—Black, with cinereous hairs.
Clypeus yellowish white. Eyes livid. Mandibles with a whitish
mark near the base. Antennæ rather shorter than the body. Thorax
and pectus thickly clothed with long hairs. Abdomen with bands of
cinereous pubescence on the hind borders of the segments; hind
borders of the ventral-segments piceous. Middle femora with gilded
tomentum beneath; tarsi tawny beneath. Wings cinereous; veins
black. Length of the body 6 lines.

A specimen from Mount Sinai is probably the female of this
species.
Wady Ferran.

262. EUCERA AMPLA. *Female.*—Black, very thickly covered with ochraceous hairs. Head above with brown hairs; clypeus testaceous and with pale hairs towards the mouth. Eyes piceous. Abdomen with black hairs towards the tip; under side with brown hairs. Legs thickly clothed with short ochraceous hairs; hind tibiæ and hind tarsi reddish. Wings cinereous; veins black. Length of the body 7½ lines.

Cairo.

263. EUCERA PILOSA. *Male.*—Black. Body, femora and fore tibiæ densely clothed with cinereous hairs. Clypeus yellow. Eyes piceous. Antennæ rather shorter than the body. Tarsi and four posterior tibiæ with shorter hairs; tarsi tawny beneath. Wings slightly cinereous; veins black. Length of the body 6 lines.

Cairo.

Genus TETRALONIA, *Spinola.*

264. TETRALONIA RUFICOLLIS. Macrocera ruficollis, *Brullé, Exp. Sci. Morée,* iii. 333, pl. 48, f. 5.

Cairo.

Inhabits Greece and North Africa.

265. TETRALONIA BLANDA. *Female.*—Black. Head and pectus with cinereous pubescence. Clypeus purple. Eyes blackish. Mouth piceous. Antennæ red. Thorax thickly covered with ochraceous pubescence. Abdomen tawny, with a dorsal band of gilded pubescence on the fore part of each segment. Legs tawny, with short testaceous hairs. Wings slightly cinereous; veins and stigma black. Length of the body 5 lines.

Harkeko.

266. TETRALONIA VETUSTA. *Male.*—Black, thickly clothed with short hoary hairs. Eyes and mouth tawny. Antennæ rather more than half the length of the body. Abdomen with a band of gilded tomentum on each segment. Legs tawny; femora not pilose; fore femora and fore tibiæ black. Wings cinereous; veins black, tawny at the base. Length of the body 4½ lines.

Most allied to T. Floralia, *Smith.*

Harkeko.

267. TETRALONIA INVARIA. *Male.*—Black. Head and thorax thickly covered with cinereous hairs. Eyes livid. Mouth black; tongue tawny. Antennæ a little shorter than the body; tawny beneath. Abdomen with testaceous tomentum and towards the base with cinereous hairs. Legs slightly pilose; tarsi tawny, excepting the first joint. Wings cinereous; veins and stigma black. Length of the body 4½ lines.

It has some affinity to T. Malvæ; the bands of the abdomen are broader and join the hind borders of the segments.

Cairo.

268. TETRALONIA AMŒNA. *Female.*—Black. Head and pectus clothed with whitish hairs. Clypeus pale luteous. Antennæ red. Thorax and legs with ochraceous hairs. Abdomen ferruginous, with three white bands which are slightly narrower in the middle, and with two posterior slightly gilded white bands. Wings dark cinereous, with a lurid tinge; veins and stigma black. *Var. β.*—Abdomen piceous, tawny towards the tip. Length of the body 6 lines.

It has most affinity to T. tricincta, a North African species, but is distinguished by the colour of the clypeus.

Rafla.

269. TETRALONIA DECORA. *Female.*—Black, stout. Head and pectus clothed with whitish hairs. Eyes ferruginous. Antennæ dark red beneath. Thorax and legs clothed with ochraceous hairs. Abdomen above with white hairs at the base and with three bands of white tomentum; second band narrower in the middle; third deeply notched in the middle of the hind side; a white subapical patch on each side; under side with tawny bands. Wings cinereous; veins and stigma black. Length of the body 5½ lines.

This is also nearly allied to T. tricincta.

Hor Tamanib.

270. TETRALONIA SPOLIATA. *Female.*—Black. Head, thorax and legs clothed with cinereous hairs. Eyes piceous. Abdomen with four dorsal bands of whitish tomentum on the hind borders of the segments; first band narrower than the others. Legs partly clothed with ochraceous hairs. Wings dark cinereous; veins and stigma black. Length of the body 7—7½ lines.

Mount Sinai.

Genus ANTHOPHORA, *Latr.*

271. ANTHOPHORA NIGROCINCTA, *St. Farg. Hym.* ii. 76.
Cairo. Heliopolis.

272. ANTHOPHORA SENESCENS, *St. Farg. Hym.* ii. 71.
Cairo.

273. ANTHOPHORA BASALIS, *Smith, Cat. Hym.* ii. 335.
Wâdy Gennèh. Wâdy Ferran.
Inhabits South Africa.

274. ANTHOPHORA CALENS, *St. Farg. Hym.* ii. 49.
Hor Tamanib. Massowah.

275. ANTHOPHORA DUBIA, *Smith, Cat. Hym.* ii. 323.
Cairo. Heliopolis.
Inhabits South France.

276. ANTHOPHORA ANNULIFERA. *Female.*—Black. Head, pectus and legs with white hairs. Head above yellow, excepting the vertex. Eyes lurid. Mouth tawny. Antennæ dark red beneath; first joint pale testaceous beneath. Thorax with yellowish hairs. Abdomen with a band of white pubescence on the hind border of each segment. Legs with black hairs beneath. Wings cinereous; veins black. Length of the body 5 lines.

A. calens, *var.*?
Wâdy Ferran.

277. ANTHOPHORA SENILIS. *Male.*—Black, stout. Head, pectus, abdomen and legs clothed with white hairs. Face luteous; clypeus yellow. First joint of the antennæ yellow beneath. Thorax clothed with lutescent cinereous hairs. Abdomen at the base clothed occasionally with cinereous hairs. Tarsi tawny, black towards the tips. Wings cinereous; veins black. Length of the body 7 lines.

Most allied to A. fulvitarsis, *Brullé;* differs rather from that species in the bands of the abdomen.
Cairo.

278. ANTHOPHORA BIMACULIFERA. *Female.* — Piceous. Head
clothed with whitish hairs and on the vertex with luteous hairs ; face
testaceous, with a large subquadrate piceous spot on each side ; clypeus
yellow. Antennæ red beneath. Thorax thickly clothed with ochraceous pubescence. Pectus with white pubescence. Abdomen with
thin cinereous pubescence ; a testaceous slightly gilded band on the
hind border of each segment ; under side and legs red. Fore legs
with whitish hairs ; four posterior legs with ochraceous hairs, their
tarsi with brown hairs. Wings cinereous ; veins black, tawny at the
base. Length of the body 6 lines.

Harkeko.

279. ANTHOPHORA MELALEUCA. *Female.*—Black. Head, thorax,
pectus and legs thickly clothed with snow-white hairs. Eyes reddish.
First joint of the antennæ whitish beneath. Abdomen with snow-
white pubescence and with four deep black bands, which from the first
to the fourth are successively narrower. Legs clothed with black
hairs beneath. Wings pellucid ; veins black, testaceous at the base.
Length of the body 4½ lines.

Cairo.

280. ANTHOPHORA LUTESCENS. *Female.* — Black, clothed with
luteous hairs. Hind part of the head with cinereous tomentum.
Face and clypeus yellow. Eyes lurid. Antennæ piceous, red beneath ;
first joint piceous, whitish beneath. Pectus with hoary hairs.
Abdomen with a testaceous band along the hind border of each
segment. Four posterior tibiæ densely pilose. Wings cinereous ;
veins black, tawny towards the base. Length of the body 4 lines.

Wâdy Ferran.

281. ANTHOPHORA CANA. *Male.*—Black. Head, thorax, pectus
and legs thickly clothed with snow-white hairs. Face and clypeus
pale testaceous. Eyes red. Antennæ piceous beneath ; first joint
yellow beneath. Abdomen with white pubescence ; hind borders of
the segments pale yellowish, this hue nearly hidden by the down.
Legs clothed with black hairs beneath. Wings pellucid ; veins
testaceous, black towards the tips. *Female.*—Face and clypeus pale
yellow. Antennæ reddish ; first joint piceous. Length of the body
4—4½ lines.

Massowah. Harkeko.

282. ANTHOPHORA PAUPERATA. *Female.*—Black. Head, thorax, pectus and legs with cinereous hairs. Eyes piceous. Mouth tawny. Antennæ tawny beneath. Abdomen with a hoary testaceous-bordered pubescent band on each dorsal segment. Legs with black hairs beneath. Wings cinereous; veins black. Length of the body 5 lines.

Mount Sinai.

283. ANTHOPHORA ILLEPIDA. *Female.*—Black. Head, thorax, pectus and legs with cinereous hairs. Face and clypeus prominent, forming a transverse ridge. Eyes piceous. Abdomen with a pale luteous band on the hind border of each segment. Tibiæ and tarsi partly clothed with luteous hairs. Wings cinereous; veins black; tegulæ tawny, large. Length of the body 6 lines.

Wâdy Ferran.

284. ANTHOPHORA BIMACULIFERA. *Female.* — Black. Head, pectus, tibiæ and tarsi with cinereous hairs. Face, clypeus and mandibles yellow; face with a large subquadrate black spot on each side; mandibles with black tips. Eyes ferruginous. Antennæ dark red, black towards the base; first joint yellow beneath. Thorax with ochraceous hairs. Abdomen with a band of cinereous tomentum along the hind border of each dorsal segment; hind borders of the ventral segments tawny. Tibiæ and tarsi with black hairs beneath. Wings cinereous; veins black. Length of the body 6 lines.

Harkeko.

285. ANTHOPHORA PUNCTIFRONS. *Female.* — Black. Head, pectus, tibiæ and tarsi clothed with white hairs. Face, clypeus and mandibles yellow; face with a black point on each side; mandibles with black tips. Mouth tawny. Eyes reddish. Antennæ red; first joint yellow beneath. Thorax with ochraceous hairs. Abdomen with a band of white pubescence on the hind border of each dorsal segment. Tibiæ beneath and tarsi with black hairs. Wings cinereous; veins black. Length of the body 5 lines.

A. bimaculifera, *var.?*

Massowah.

286. ANTHOPHORA PULVEREA. *Female.*—Black. Head, thorax, pectus and the tibiæ and tarsi of the four anterior legs with white hairs. Face and clypeus yellow, the former divided from the latter

by a slight transverse ferruginous ridge. Mouth ferruginous. Eyes red. Antennæ red beneath. Abdomen with a broad band of white pubescence along the hind border of each dorsal segment. Hind tibiæ beneath and hind tarsi with black hairs, the former with bright ochraceous hairs above. Wings cinereous ; veins black. *Var. β.*— Hind tibiæ with yellowish white hairs above. Length of the body 5 lines.

Harkeko. Wády Hebran.

287. Anthophora inclyta. *Female.*—Black, clothed with bright ochraceous hairs. Head beneath with cinereous hairs ; face and clypeus not hairy ; fore border of the face forming a transverse ridge. Eyes piceous. Mouth tawny. Legs with hairs like those of the body in colour; hind tibiæ and hind tarsi very densely pilose. Wings cinereous; veins black. *Var. β.*—Pectus with cinereous hairs. Length of the body 7½—8 lines.

Most allied to A. pennata, but is larger and brighter and has darker wings.

Rafla. Wády Ferran. Mount Sinai.

288. Anthophora superans. *Female.*—Vertex, thorax, pectus and base of the abdomen thickly clothed with pale ochraceous hairs. Head excepting the vertex white and without hairs. Eyes and mouth ferruginous. First joint of the antennæ white beneath. Abdomen with three bands of pale ochraceous pubescence on the hind borders of the segments. Legs fringed with cinereous hairs ; tibiæ and tarsi clothed with cinereous pubescence; tarsi with ochraceous hairs beneath. Wings slightly cinereous ; veins black. Length of the body 7½ lines.

Mount Sinai.

Genus XYLOCOPA, *Latr.*

289. Xylocopa æstuans. Apis æstuans, *Linn. Syst. Nat.* i. 961.

Cairo. Heliopolis. Harkeko. Wády Gennèh.

Inhabits Hindostan.

290. Xylocopa olivacea. Apis olivacea, *Fabr. Ent. Syst.* ii. 319.

Hor Tamanib.

291. XYLOCOPA LANATA, *Smith, Cat. Hym.* ii. 345.
Wâdy Gennêh. Mount Sinai.
Inhabits Turkey.

292. XYLOCOPA VIOLACEA. Apis violacea, *Linn. Syst. Nat.* i. 959.
Hor Tamanib. Massowah. Harkeko.
Inhabits South Europe.

Genus APIS, *Linn.*

293. APIS FASCIATA, *Latr. Ann. Mus. Hist. Nat.* v. 171, pl. 13,
f. 9.
Cairo. Heliopolis.

ERRATUM.

Page 18.
For EBERRINA *read* EBENINA.